AF333825

Springer Series in

OPTICAL SCIENCES

Y. Aizu   T. Asakura

Spatial Filtering Velocimetry

Fundamentals and Applications

With 112 Figures

 Springer

Professor Yoshihisa Aizu
Muroran Institute of Technology, Department of Mechanical Systems Engineering
Mizumoto 27-1, Muroran 050-8585, Hokkaido, Japan
E-mail: aizu@mmm.muroran-it.ac.jp

Professor Toshimitsu Asakura
President, Hokkai-Gakuen University
Asahi-Machi 4-1-40, Toyohira-Ku, Sapporo 062-8605, Hokkaido, Japan
E-mail: asakura@hgu.jp

ISSN 0342-4111

ISBN-10 3-540-28186-X Springer Berlin Heidelberg New York

ISBN-13 978-3-540-28186-3 Springer Berlin Heidelberg New York

Library of Congress Control Number: 2005933899

Springer is a part of Springer Science+Business Media.

springeronline.com

© Springer-Verlag Berlin Heidelberg 2006
Printed in The Netherlands

Typesetting: SPI, Pondicherry, India
Cover concept by eStudio Calamar Steinen using a background picture from The Optics Project. Courtesy of John T. Foley, Professor, Department of Physics and Astronomy, Mississippi State University, USA.
Cover production: *design & production* GmbH, Heidelberg

Printed on acid-free paper   SPIN: 11531180   57/3141/SPI   5 4 3 2 1 0

To our families

Preface

The invention of lasers in the early 1960s enhanced the rapid development of optoelectronics which had introduced various optical measurement methods. A typical example of the methods is found in measurements of velocity. It is well recognized that optical velocity measuring methods have important advantages, such as noncontacting and nondisturbing operations, over conventional methods employed previously. These fundamental advantages are indicated by the enormous research effort which has gone into their development for many years. One of the optical methods proposed and studied to measure the velocity is laser Doppler velocimetry which was proposed in the early 1960s and extensively studied by many investigators and is at present applied to practical uses. Another is spatial filtering velocimetry which was also proposed in the early 1960s and studied by a number of investigators. In comparison with laser Doppler velocimetry, spatial filtering velocimetry had not received much attention from investigators but was studied steadily by several research groups mainly in Japan and is now practically used in various fields of engineering.

Several important books on laser Doppler velocimetry have already been published, but there has been no book on spatial filtering velocimetry. This book is the first contribution to spatial filtering velocimetry. Therefore, the Introduction of Chapter 1 provides in detail a historical review of spatial filtering velocimetry, relating it to other optical methods and discussing its practical relevance. In the book following Chap. 1, the most important results on the subject both from our own papers and from publications by other authors have been collected together and presented in a concise and easily readable form. Special emphasis has been placed on the fundamental content of spatial filtering velocimetry in a general form and on a wide range of systems and applications of this velocimetry. By following this emphasis, the contents of this book consisting of six chapters may be divided into two main parts: the fundamentals given in Chaps. 2–4 and the systems and applications in Chaps. 5 and 6.

Since the subject matter of this book is interdisciplinary, we have tried to make the book self-contained and easily understandable to readers in various fields. Considering the wide-ranging backgrounds of readers, we have also attempted to give a comprehensive list of the most pertinent references at the end of the book.

The authors wish to express their thanks to Prof. A. Kobayashi formerly at the Tokyo Institute of Technology for kindly supplying reprints of his papers on spatial filtering velocimetry, which were very helpful in the preparation of fundamental parts in this book. We are also grateful to several researchers who provided us useful information on their works in spatial filtering velocimetry, especially Dr. K. Michel at the Jena-Optronik GmbH, Prof. S.G. Hanson at the Risø National Laboratory, and Prof. K. Oka at Hokkaido University. Y. Aizu is thankful to Prof. Y. Itakura at Shiga University for kind correspondence which began with discussions on spatial filtering velocimetry. Finally, we would like to thank Mr. T. Ushizaka for his valuable cooperation in various studies on spatial filtering velocimetry which were performed formerly by our group at Hokkaido University.

Muroran and Sapporo, Hokkaido,   *Yoshihisa Aizu*
May 2005   *Toshimitsu Asakura*

Contents

1

Introduction

1.1 Survey of Optical Velocimetry

The development of optics and electronics has established the great importance of optical metrology in various fields of science and engineering. In particular, the invention of high-intensity coherent light sources known as lasers has introduced many optical measurement techniques that had previously been unavailable. A typical example of such techniques is found in velocity measurements. In comparison with conventional velocity measurement techniques such as pitot tubes and hot-wire devices, optical techniques have important and practical advantages of noncontacting and nondisturbing operations. Therefore, a large amount of research effort has been made to develop techniques for practical uses in the world. Today, there are many commercial instruments provided in the market, which are widely used in laboratory and industrial applications, for example, fluid mechanics, aerosol science, biomedical engineering, and remote sensing.

A variety of optical techniques have been proposed and studied for measurements of velocity. For convenience of a survey, let us divide them into two types incoherent and coherent techniques. Note that these types do not necessarily mean the use of incoherent or coherent light sources. The incoherent technique uses information on the light intensity of the image of an object, whereas the coherent technique uses information on the amplitude and phase of the light. The difference between these types is generally recognized by *"images"* and *"interference patterns."* Some representative techniques categorized in the two types are concisely reviewed in the literature [1]. The type of incoherent techniques contains photography and cinematography used in the early stages of the development of optical velocimetry. Simply observing or photographing the motion path of an object is primitive but an easy and convenient means for measuring velocity. A photograph recording the trace of a particle image allows us to determine the velocity with knowledge of both the exposure time and the trace length. Alternatively, the velocity can be determined by measuring the time for the passing of a particle image through a window area with known dimensions. This scheme may be realized by replacing the photographic film and the measuring scale with a photodetector having the detecting window and an oscilloscope, which records the temporal signal of the light intensity of images passing through the window. A similar method to this approach uses two detecting slits with a known separation instead of the detecting window. In this case, the velocity is determined by measuring the transit time of particle images across the two slits. This method is also realized by using two laser beams focused in the object plane with a known separation and removing the two slits. This technique is referred to as the time-of-flight method or the two-beam cross-correlation method [2]. By following the method using two slits in the image plane, it is natural to analogize the use of multiple slits since it sequentially gives multiple data of the transit time, and, thus, improvement of the measurement accuracy is expected. For example, passage of particle images on periodically arrayed parallel slits yields electric signals with periodic variations of light intensity. The instantaneous velocity of particles is determined by measuring the frequency of the variation with a known pitch of slits. This approach was historically referred to as the photoelectric image-tracing technique [1] and is currently called the spatial filtering technique, which is the subject of this book.

Cinematography records images of moving particles on cine films and, then, particle positions on successive frames, and the time interval between the frames allows the velocity to be determined. This technique provides us with a two-dimensional velocity distribution if each of the particle images on the frames is analyzed individually. In principle, the velocity is obtained from just two successive frames. A double-exposure photograph obtained by using two successive laser pulses records two instantaneous distributions of particle images at two successive instants. Each pair of particle images is used to analyze the moving direction and velocity of each particle with the known time interval of pulses, and, thus, analyzing the whole pairs on the photograph gives a two-dimensional visualization of the velocity-vector distribution. This analysis involves finding the motion loci of a large number of particles, which becomes possible on a practical level with the aid of a fast computer. This technique is currently known as particle image velocimetry (PIV) and is widely used in the fields of fluid dynamics and aerosol science [3]. In measurements of flow systems with refractive-index gradients, shadowgraphy and schlieren are familiar optical methods. Although these methods give us the two-dimensional distribution of a shadow which is related to the flow condition, the velocity is obtained indirectly.

The coherent technique uses the principle of interferometry, which is realized by interference of mutually coherent light beams. Generally, interference is observed in the forms of fringe patterns or speckle patterns. The interference fringes are used to determine the particle velocity by observing the transit time of a particle moving across the fringes. This concept is well known as laser Doppler velocimetry (LDV) [1, 2, 4–6], which is realized in some different ways and widely used in various fields of science and engineering. In a typical model of the LDV, a light beam from a laser source is split into two beams, which cross in their focused position to form an interference fringe pattern in a probing volume. Since the beams are generally focused to several to several tens of microns in diameter, LDV measurements enable us to determine the local velocity in a small volume having such a dimension. This high spatial resolution is a very powerful advantage of the LDV technique. When a scattering object is, for example, a diffusing plate or rough surface, the scattered light forms a speckle pattern as a result of the interference of many scattered waves having mutually random phases. Motion of the object causes the temporal and spatial variations of interference intensity in the speckle pattern, and, thus, the velocity information can be extracted. Frequency or correlation analysis is applied to photoelectric signals of the speckle intensity variation, and the velocity of the object can be determined. This technique is referred to as laser speckle velocimetry (LSV) [7]. Although there are different ways for frequency analysis in LSV, one of them uses the spatial filtering technique described in this book. The speckle pattern recorded in an image plane potentially contains the two-dimensional velocity information of an object plane. Thus, photographic or computer image analysis is applied to the speckle pattern to visualize the velocity distribution. This approach may be regarded as velocimetric application of laser speckle photography [8].

Among the various above-mentioned incoherent and coherent techniques, the laser Doppler method has been most extensively studied by a number of researchers because of its high spatial resolution and high measurement accuracy. On the other hand, the spatial filtering method has not received much attention from investigators at the beginning stage of research, although its measurement performance is similar to that of the Doppler method. In addition to this, the spatial filtering method has practical advantages such as simplicity and stability of optical and mechanical systems and a choice of light sources. By appreciating the value of these features, the spatial filtering method has been studied for the development of practical instruments as well as laser Doppler velocimeters.

To the authors' knowledge, the basic concept that leads to spatial filtering velocimetry was developed from aerial camera-control techniques [9] and infrared optical tracking techniques [10,11] which used gratings or reticles. The definite proposal of the spatial filtering method for velocity measurements is found in the study of Ator [12]. He showed conceptually the principle of the method by simulating the operation of a parallel-slit reticle as the spatial filter. Ator [13] also provided the interpretation of the method from the standpoint of correlation theory. The first experimental demonstration of the method was made by Gaster [14] for the study of fluid flows. The theoretical basis of the method was given by Naito et al. [15] and Tsutsumi [16], separately. As an example of the spatial filter, they treated a transmission grating and analyzed the power spectral density function of its transmittance in a space domain. The results successfully demonstrated that the transmission grating acted as a spatial filter which was available for velocity measurements. To improve the selectivity of the spatial filter, Kobayashi and Naito [17] discussed the optimizing problem of a narrow-band-pass spatial filter. On the basis of these fundamental studies, the group of Kobayashi [18,19] developed the spatial filtering detector which was a photodetector having the function of a spatial filter. Tsutsumi [20] studied theoretically the filtering characteristics of a parallel-slit reticle by developing conventional filtering theory in the time domain. According to these fundamental characteristics, Itakura et al. [21] constructed a new type of spatial filter using a liquid crystal cell array, which enabled two-dimensional velocity components to be measured. To improve the filtering characteristics in a lower spatial frequency region, Tsudagawa et al. [22] modified the spatial filter by introducing the field-of-view of a parallelogram. Asakura and his co-workers [23–28] extensively studied the method using a transmission grating based on Gaster's proposal. Ushizaka and Asakura [23] developed an optical imaging system with a microscope for spatial filtering velocimetry and applied the system to measurements of the flow velocity distribution in small glass tubes having diameters of $130\,\mu$m to 3.5 mm. Aizu et al. [25] constructed a differential-type transmission grating velocimeter to improve the ability of removing undesirable low-frequency components and showed its usefulness for measurements of flow velocity in microscopic regions. This type of spatial filtering velocimeter was used by some researchers separately, to measure blood flow velocity: Koyama et al. [24], Aizu et al. [28], Borders and Granger [29], and Reuter and Kratzer [30]. Using this method, Delatour and Hanss [31] measured the electrophoretic mobility distribution.

In addition to the transmission grating or parallel slit reticle, other optical elements are also available for spatial filters. Hayashi and Kitagawa [32–35] constructed a novel spatial filter with an optical fiber array and applied to measurements of two-dimensional velocity components and distance with directional determination. Mitsuhashi and Mochizuki [36] studied a type of spatial filtering method using an image sensor by which the spatial filter was electronically realized. The prism grating is an interesting example of optical elements that are available for spatial filters. Using this, researchers such as Röckemann and Plesse [37], Slaaf et al. [38,39], Reuter and Talukder [40], and Kiesewetter et al. [41] measured blood flow velocity. Ushizaka et al. [42] studied the imaging and deflection properties of light rays of a lenticular grating and showed that it acted as a spatial filter similar in principle to the prism grating.

The principle of spatial filtering velocimetry is applied in different ways to measurements of moving objects and their motion. With an analogy to the dual-beam LDV system, Ballik and Chan [43–45] studied theoretically and experimentally a fringe image technique in which a grating-like illumination was made on a moving object and the intensity-modulated scattered light was received by a photodetector without a grating in front. Aizu et al. [46] developed the spatial filtering method so that it enabled us to determine a velocity gradient. Using the spatial filtering detector, Ohno et al. [47] proposed a method for measurements of random motion in two dimensions, such as the average velocity, number, and size of moving objects. Kobayashi et al. [48]

investigated the fluctuation of a rotating velocity obtained by using a spatial filtering detector. Wang and Tichenor [49] proposed a new type of transmission grating constructed with gradually varying pitches, which enables the diameter of a moving particle to be measured. Based on optical imaging properties, the spatial filtering method was also applied to measurements of focusing errors [16, 50], distance, and displacement [51, 52]. There are other studies of spatial filtering devices and systems, and their applications, which will be described in this book.

1.2 Spatial Filtering Velocimetry

The principle of spatial filtering velocimetry (SFV) will be described by the narrow-band-pass spatial filtering effect of spatially periodic transmittance. However, it may be readily and intuitively understood from the light intensity modulation of a moving image caused by the transmittance. Figures 1.1 and 1.2 show schematically the basic optical system and the principle of spatial filtering velocimetry. The basic operation of this velocimetry is to observe the optical image of a moving object through the spatial filter such as a set of parallel slits or a transmission grating. The illuminating light is scattered by an object, such as a small particle, moving with velocity v_0 in the direction x_0. A lens L forms an image of the object onto a spatial filter (SF) that has spatially periodic transmittance in the moving direction (x axis) of the object, or onto grating lines set perpendicularly to the direction of motion. The light passing through the spatial filter is received by a photodetector (PD). The total intensity of the light detected by the PD varies periodically because of the image movement with constant velocity v and the periodic transmittance having a pitch p, as shown in Fig. 1.2. Thus, the output from the PD provides a periodic signal containing a period $T_0 = p/v$. By measuring the frequency $f_0 = 1/T_0$ of this signal, the object velocity v_0 is determined from

$$v_0 = \frac{p}{M} f_0 , \tag{1.1}$$

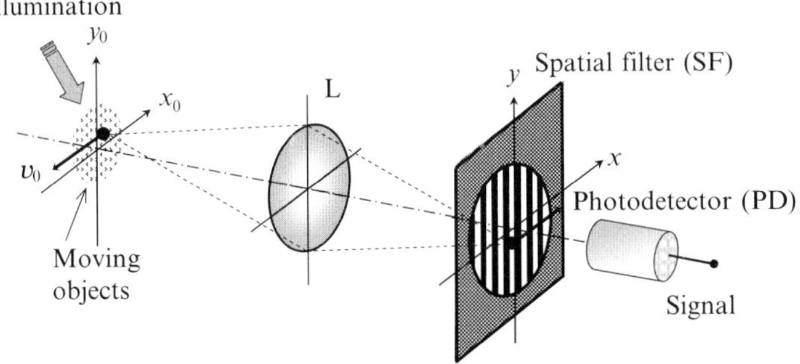

Fig. 1.1. Basic optical system of spatial filtering velocimetry

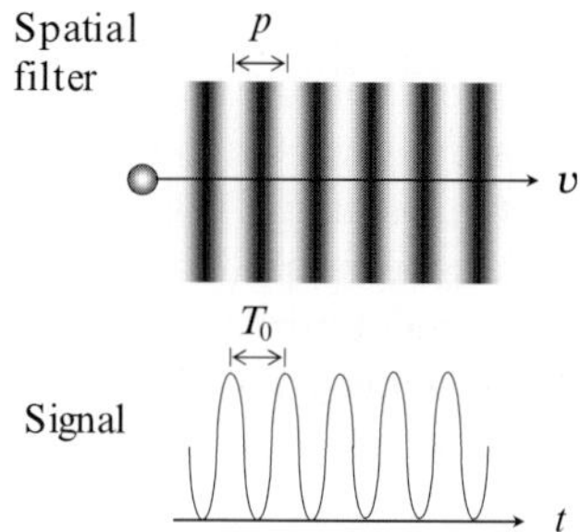

Fig. 1.2. Principle of spatial filtering velocimetry

where M is the optical magnification of the imaging system with lens L and, thus, $v = Mv_0$. As depicted in Fig. 1.2, the output signal has a periodic waveform characterized by frequency f_0, which is usually sinusoidal, and then the frequency is simply measured by a frequency counter or a spectrum analyzer, at least in principle. Thus, a simple construction of a velocity-measuring system can be realized by spatial filtering velocimetry.

The principle of spatial filtering velocimetry may be compared with that of differential-type LDV. As illustrated in Fig. 1.3, the operation of this type of LDV can be understood in terms of interference fringes that exist in the cross-over region of two illuminating beams. The pitch p_d of these fringes is given by the equation [5]

$$p_\mathrm{d} = \frac{\lambda}{2 \sin \dfrac{\alpha}{2}} \, , \qquad (1.2)$$

where λ is the wavelength of the illuminating laser light and α is the angle between the two beams. When a particle crosses light and dark fringes perpendicularly with velocity v_0 in that region, modulation of the intensity at the photodetector arises from the variation in the light illuminating the particle. The frequency f_d of this modulation is, thus, given by

$$f_\mathrm{d} = \frac{v_0}{p_\mathrm{d}} \, , \qquad (1.3)$$

and the velocity v_0 is determined by the relation

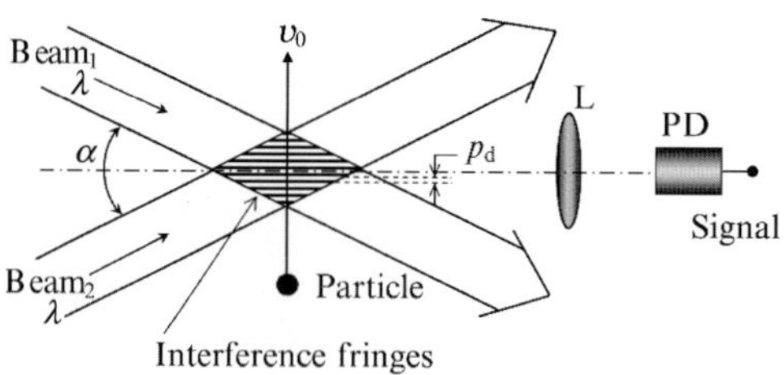

Fig. 1.3. Basic configuration of the differential-type LDV

$$v_0 = f_\mathrm{d}\, p_\mathrm{d} = \frac{\lambda}{2 \sin \dfrac{\alpha}{2}}\, f_\mathrm{d} \,. \tag{1.4}$$

In both velocimetric methods of spatial filtering and differential LDV, a combination of the moving scattering object having a constant velocity and the spatial periodicity of a transmission grating or interference fringes produces periodic output signals. This similarity may help in understanding signal characteristics, signal analysis, system design, and applications of the spatial filtering method through this book.

1.3 The Book

The intention of this book is to present the fundamental content of spatial filtering velocimetry in a general form and to review a wide range of systems and applications of this velocimetry. The book consists of six chapters, which may be divided into two main parts: the fundamentals described in Chaps. 2–4 and the systems and applications in Chaps. 5 and 6.

Chapter 2 introduces the principle and the fundamental properties of spatial filtering velocimetry in theoretical expressions. Here, a transmission grating having parallel slits is considered a typical example of the spatial filter. The theory will be developed on the basis of narrow-band-pass filtering characteristics in the space domain in relation to geometric parameters of the spatial filter. The effects of various scattering objects on the filtering performance are also discussed for consideration of the basic accuracy in this velocimetry.

Chapter 3 reviews some key matters in the framework of the general imaging theory, which are useful for interpreting and designing the optical system of spatial filtering velocimetry. They include resolution, transfer function, and lens aberrations. Imaging properties are also discussed in connection with the focusing depth, probe volume, and illumination which are rather unique to this velocimetry. In Chap. 4, signal-analyzing techniques are discussed for reliable determination of the central frequency of photodetector outputs. Since the velocity of objects is determined directly from the frequency measurement, the signal analysis should be made with better accuracy in measurement. For this purpose, the chapter briefly introduces various techniques available for spatial filtering velocimetry and gives some guidelines for choosing the appropriate technique under given conditions.

Spatial filtering velocimeters are constructed with a spatial filtering device as the key component, together with proper optical and signal-analyzing systems. Then, the performance and specification of instrumental systems depend on the spatial filtering device being employed. Chapter 5 introduces a variety of spatial filtering devices and presents the corresponding instrumental systems using such devices. For better understanding, the devices and systems are divided into eight categories in this book according to differences in the operation of spatial filtering. They are briefly summarized and compared in the last section of this chapter.

Chapter 6 is devoted to various applications of spatial filtering velocimetry. The well-known LDV technique is generally applicable to local-velocity measurements because of the use of focused laser beams. As contrasted with this, the spatial filtering technique has a wide range of scaling flexibility for objects being measured, for instance, from microscopic flows to large aircraft. The simplicity of the measuring principle and configuration is another key factor that supports applicability. This chapter outlines interesting examples of these applications ranging from laboratory experiments to industrial measurements. Spatial filtering velocimetry can be used for measurements of some physical quantities besides velocity, which are also briefly included with other related velocity-measuring techniques in the last sections of this chapter.

2

Principle and Properties of the Spatial Filtering Method

The basic operation of spatial filtering velocimetry (SFV) is observing the optical image of a moving object such as a small particle through a set of parallel slits or a transmission grating. When the light from the moving image passes through the grating, it acts as a narrow-band-pass spatial filter which picks up particular spatial frequency components from the image intensity distribution. Therefore, the principle of the spatial filtering method is mathematically described by power spectral density functions of the image and the grating in the spatial frequency domain. Since the intuitive operation in SFV can simply be interpreted by (1.1), the mathematical treatment of spatial filtering might be considered unessential. However, the signal quality and measurement accuracy depend directly on both the filtering characteristics of the grating and the spatial distribution of the image intensity. Thus, the theoretical background of the spatial filtering method is necessary for the design of the spatial filter, the configuration of the optical system, and application to various objects.

In this chapter, the principle and fundamental properties of the spatial filtering method are theoretically described. In Sect. 2.1, the spatial filtering effect of a grating is given in terms of the convolution integral in the spatial frequency domain. Section 2.2 theoretically reveals that a spatially periodic transmittance realizes narrow-band-pass filtering in the space domain. Quantitative descriptions of the spatial filtering effect and its characteristics are presented in Sects. 2.3 and 2.4 based on power spectral analysis for some possible transmission gratings. Section 2.5 gives briefly a design program of the spatial filter for realizing desirable filtering characteristics. In Sect. 2.6, effects of scattering objects on output signals are discussed from the standpoint of particle sizes. Finally, properties required for objects being measured in SFV are discussed in three examples of small particles, rough surfaces, and a speckle pattern in Sect. 2.7.

2.1 Spatial Filtering Effect

In this section, a brief conception is described for the spatial filtering effect
on the image intensity distribution. Figure 2.1 schematically shows the basic
optical system of the spatial filtering method for velocity measurements. The
image plane in which a spatial filter is placed is designated by a coordinate
system (x, y), and the direction of the transmitted light is denoted by the z
axis. The image quality is one of the factors that characterize output signals.
For simple treatments, however, it is assumed here that an ideal image is
formed on the spatial filter. The influences of image quality on signals will be
discussed in Sect. 2.7. Let $f(x, y)$ and $h(x, y)$ be the light intensity distribution
of the moving image in the x–y plane and the light intensity transmittance of
the spatial filter, respectively. All light passing through the spatial filter is
assumed to be received by the photodetector. When the image moves with
velocity components v_x and v_y in directions x and y, respectively, the output
signal $g(x_r, y_r)$ obtained from the photodetector is given by the convolution
integral (Appendix A.4) as

$$g\left(x_{\mathrm{r}}, y_{\mathrm{r}}\right) = \iint_{-\infty}^{\infty} f\left(x_{\mathrm{r}} - x, y_{\mathrm{r}} - y\right) h\left(x, y\right) \, \mathrm{d}x \, \mathrm{d}y \;, \qquad (2.1)$$

where $x_{\mathrm{r}} = v_x t + c_1$, $y_{\mathrm{r}} = v_y t + c_2$, and c_1 and c_2 are constants. The function
$h(x, y)$ in (2.1) corresponds to the impulse response in electric communica-
tion theory. Generally, the image intensity distribution $f(x, y)$ is considered
to follow a random process in time and space. A typical example is imaging of
scattering particles randomly distributed in a measuring volume. Then, a statistical treatment is introduced for $f(x, y)$ in the following analysis. With the
assumption that the intensity distribution $f(x, y)$ follows a stationary, random
ergodic process in two dimensions, the autocorrelation function $R(\tau_x, \tau_y)$ of

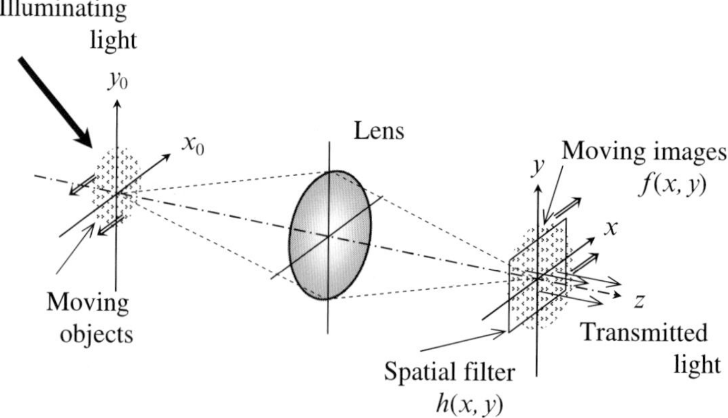

Fig. 2.1. Basic optical system of the spatial filtering method for velocity measurements

the signal is written as

$$R\left(\tau_{\mathrm{x}}, \tau_{\mathrm{y}}\right) = \mathbf{E}\left[g\left(x_{\mathrm{r}} + \tau_{\mathrm{x}}, y_{\mathrm{r}} + \tau_{\mathrm{y}}\right) g^{*}\left(x_{\mathrm{r}}, y_{\mathrm{r}}\right)\right] , \qquad (2.2)$$

where the notation $\mathbf{E}[\cdots]$ means the expectation operation [53]. Apart from a constant factor, the spatial power spectral density function $G_{\mathrm{p}}(\mu, \nu)$ of the function $g(x_{\mathrm{r}}, y_{\mathrm{r}})$ is derived by the Fourier transform of the correlation function $R(\tau_{\mathrm{x}}, \tau_{\mathrm{y}})$ as [15]

$$G_{\mathrm{p}}\left(\mu, \nu\right) = F_{\mathrm{p}}\left(\mu, \nu\right) H_{\mathrm{p}}\left(\mu, \nu\right) , \qquad (2.3)$$

where $F_{\mathrm{p}}(\mu, \nu)$ and $H_{\mathrm{p}}(\mu, \nu)$ stand for power spectral density functions of the image intensity $f(x, y)$ and the transmittance $h(x, y)$, respectively, both in the spatial frequency domain. These functions are expressed with $f(x, y)$ and $h(x, y)$ as follows:

$$F_{\mathrm{p}}\left(\mu, \nu\right) = \lim_{T_{\mathrm{x}}, T_{\mathrm{y}} \to \infty} \frac{1}{4 T_{\mathrm{x}} T_{\mathrm{y}}}$$
$$\cdot \left| \int_{-T_{\mathrm{x}}}^{T_{\mathrm{x}}} \int_{-T_{\mathrm{y}}}^{T_{\mathrm{y}}} f\left(x, y\right) \exp\left[-\mathrm{i} 2\pi\left(\mu x + \nu y\right)\right] \mathrm{d}x\, \mathrm{d}y \right|^{2} , \qquad (2.4)$$

$$H_{\mathrm{p}}\left(\mu, \nu\right) = \left|H\left(\mu, \nu\right)\right|^{2}$$
$$= \left| \iint_{-\infty}^{\infty} h\left(x, y\right) \exp\left[-\mathrm{i} 2\pi\left(\mu x + \nu y\right)\right] \mathrm{d}x\, \mathrm{d}y \right|^{2} , \qquad (2.5)$$

where $H(\mu, \nu)$ is the Fourier spectrum of the function $h(x, y)$ and μ and ν denote the spatial frequencies in the x and y directions, respectively.

If the image intensity distribution $f(x, y)$ is not random but periodic or nonperiodic (transient) [54], the power spectrum $F_{\mathrm{p}}(\mu, \nu)$ can be given by

$$F_{\mathrm{p}}\left(\mu, \nu\right) = \left|F\left(\mu, \nu\right)\right|^{2}$$
$$= \left| \iint_{-\infty}^{\infty} f\left(x, y\right) \exp\left[-\mathrm{i} 2\pi\left(\mu x + \nu y\right)\right] \mathrm{d}x\, \mathrm{d}y \right|^{2} , \qquad (2.6)$$

where $F(\mu, \nu)$ is the Fourier spectrum of the function $f(x, y)$. The power spectrum $G_{\mathrm{p}}(\mu, \nu)$ in this case is expressed as

$$G_{\mathrm{p}}\left(\mu, \nu\right) = \left|G\left(\mu, \nu\right)\right|^{2}$$
$$= \left|F\left(\mu, \nu\right) H\left(\mu, \nu\right)\right|^{2} = \left|F\left(\mu, \nu\right)\right|^{2} \left|H\left(\mu, \nu\right)\right|^{2} , \qquad (2.7)$$

where $G(\mu, \nu)$ is the Fourier spectrum of the output signal $g(x_{\mathrm{r}}, y_{\mathrm{r}})$, and is given by the convolution theorem as

$$G\left(\mu, \nu\right) = \iint_{-\infty}^{\infty} g\left(x_{\mathrm{r}}, y_{\mathrm{r}}\right) \exp\left[-\mathrm{i} 2\pi\left(\mu x_{\mathrm{r}} + \nu y_{\mathrm{r}}\right)\right] \mathrm{d}x_{\mathrm{r}}\, \mathrm{d}y_{\mathrm{r}} . \qquad (2.8)$$

In (2.3), the power spectrum $G_\mathrm{p}(\mu, \nu)$ is represented by multiplication of two power spectra $F_\mathrm{p}(\mu, \nu)$ and $H_\mathrm{p}(\mu, \nu)$. This relation indicates that the output signal is given by the input image modified by the spatial filter. According to linear filtering theory, (2.3) shows that $H_\mathrm{p}(\mu, \nu)$ operates as a linear filter on the input $F_\mathrm{p}(\mu, \nu)$ in the spatial frequency domain. The essentials for the expression of power spectra are given in Appendix B.

As described in the next section, the spatial filter is required to have a periodic transmittance in the moving direction of the images being measured. For mathematical simplicity, it is assumed that the image moves along the x axis with velocity $v_\mathrm{x} = v$, and $v_\mathrm{y} = 0$. In this case, the transmittance $h(x, y)$ of the spatial filter should be periodic only in the x direction and should be uniform in the y direction. The power spectral density function $G_\mathrm{p}(f)$ in the time domain is derived, by integration of (2.3) with respect to the spatial frequency ν, as

$$G_\mathrm{p}(f) = \frac{1}{v} \int_{-\infty}^{\infty} F_\mathrm{p}\left(\frac{f}{v}, \nu\right) H_\mathrm{p}\left(\frac{f}{v}, \nu\right) \mathrm{d}\nu , \qquad (2.9)$$

where the relation

$$\mu = \frac{f}{v} \qquad (2.10)$$

was used and f is the frequency in the time domain. A scaling factor $1/v$ on the right-hand side of (2.9) arises from the nature of the density function. This spectrum $G_\mathrm{p}(f)$ corresponds to that of output signals actually observed by a spectrum analyzer. The case of an image moving with a velocity component v_y in the y direction will be treated in Sect. 2.4. Equation (2.9) indicates again that the power spectrum $H_\mathrm{p}(f/v, \nu)$ works as a filtering function on the input function $F_\mathrm{p}(f/v, \nu)$. If the sizes of scattering objects such as particles are sufficiently small compared with the size of the probing volume, the power spectrum $F_\mathrm{p}(\mu, \nu)$ is considered nearly constant, or $F_\mathrm{p}(\mu, \nu) \cong 1$ under the white-noise approximation. On the other hand, the power spectrum $H_\mathrm{p}(\mu, \nu)$ of the spatial filter having a periodic transmittance in the x direction contains a narrow-band spectral component centered at the spatial frequency $\mu = \mu_0$. Figure 2.2 illustrates a typical distribution of the power spectra $F_\mathrm{p}(\mu, \nu)$ and $H_\mathrm{p}(\mu, \nu)$ at $\nu = 0$. The product of the two spectra, thus the power spectrum $G_\mathrm{p}(\mu, \nu)$ is, then, dominantly characterized by the power spectrum $H_\mathrm{p}(\mu, \nu)$, and the temporal power spectrum $G_\mathrm{p}(f)$ may contain a frequency peak at $f = f_0 = \mu_0 v$. Therefore, the image velocity may be determined by measuring the central frequency f_0 and using the relation $v = p f_0$ in (1.1). In this way, narrow-band-pass spatial filtering by periodic transmittance provides the principle of velocity measurements and, therefore, the present technique is generally called *"spatial filtering velocimetry."*

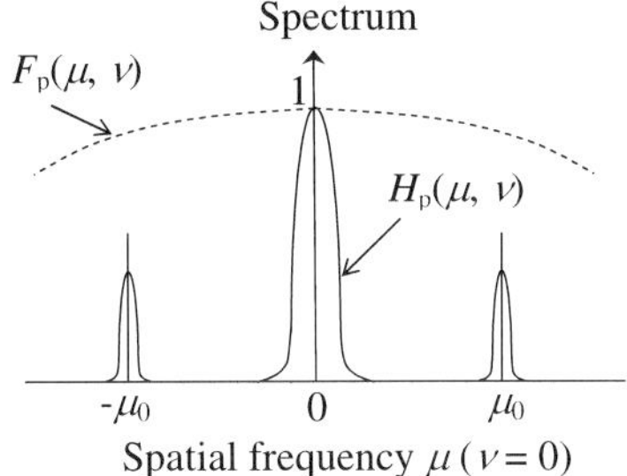

Fig. 2.2. Typical forms of power spectra $F_{\mathrm{p}}(\mu,\nu)$ and $H_{\mathrm{p}}(\mu,\nu)$ at $\nu = 0$

2.2 Transmittance Functions

To realize narrow-band-pass filtering in the space domain, the spatial filter for velocimetry is required to have a spatially periodic transmittance in the direction of object movements. By this transmittance, the moving image intensity is periodically modulated relative to the object's velocity, and periodic signals are obtained from the photodetector. A typical example of a spatial filter is a transmission grating or a set of parallel slits. This section shows how this kind of transmittance contributes to narrow-band-pass spatial filtering characteristics. Figure 2.3 shows a schematic model of a typical spatial filter and the (x, y) coordinate system. This model has periodic transmittance with a period p only in the x direction, and its area is restricted by a rectangular window having sizes X and Y in the x and y directions, respectively. The spatial filter having this form of window will be referred to as a "rectangular type" in this book. The transmittance function is given by

$$h\left(x, y\right) = h\left(x\right) = h\left(x + mp\right) , \qquad (2.11)$$

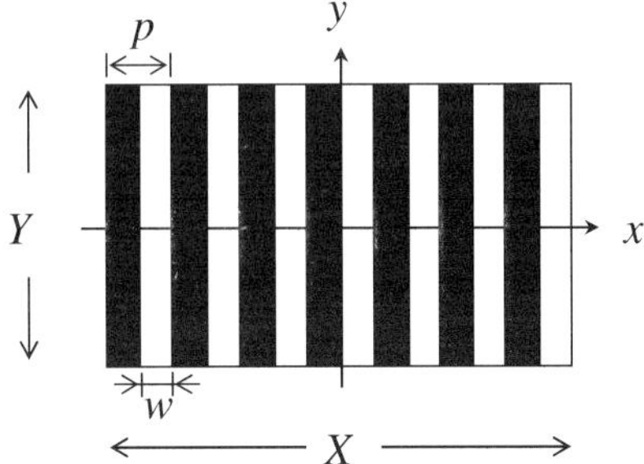

Fig. 2.3. Schematic model of a typical spatial filter (a rectangular-type transmission grating)

where m is an integer. Substitution of (2.11) in (2.5) yields the power spectrum of the transmittance. The Fourier integral becomes finite in the ranges $-X/2$ to $X/2$ and $-Y/2$ to $Y/2$, and zero otherwise. Thus, the power spectrum is derived as [15] (Appendix C)

$$|H(\mu,\nu)|^2 = X^2 Y^2 |H_\mathrm{Y}(\nu)|^2 |H_\mathrm{X}(\mu)|^2 |H_\mathrm{c}(\mu)|^2 |H_\mathrm{s}(\mu)|^2 , \qquad (2.12)$$

where

$$|H_\mathrm{Y}(\nu)|^2 = \left(\frac{\sin \pi\nu Y}{\pi\nu Y}\right)^2 , \qquad (2.13)$$

$$|H_\mathrm{X}(\mu)|^2 = \left(\frac{\sin \pi\mu X}{\pi\mu X}\right)^2 , \qquad (2.14)$$

$$|H_\mathrm{c}(\mu)|^2 = \left(\frac{\pi\mu p}{\sin \pi\mu p}\right)^2 , \qquad (2.15)$$

$$|H_\mathrm{s}(\mu)|^2 = \left|\frac{1}{p}\int_0^p h(x)\exp(-\mathrm{i}2\pi\mu x)\,\mathrm{d}x\right|^2 . \qquad (2.16)$$

In (2.12), $X^2 Y^2$ means the contribution of power due to the window area of the spatial filter. The two functions $|H_\mathrm{Y}(\nu)|^2$ and $|H_\mathrm{X}(\mu)|^2$ are given by the form $\mathrm{sinc}(x) = \sin \pi x/\pi x$, which is called the "sinc function" [55], and is derived as the Fourier transform of a rectangular function. They express, thus, the contribution of the rectangular shape which forms the window area of the spatial filter. The function $|H_\mathrm{c}(\mu)|^2$ is due to periodicity in the interval p. The last function $|H_\mathrm{s}(\mu)|^2$ is obtained from the Fourier transform of the transmittance $h(x)$ over one period and indicates the contribution of the transmittance function within one slit. The sinc function $\mathrm{sinc}(x)$ is depicted in Fig. 2.4, in which the first zero-crossing occurs at $x = \pm 1$. Then, similarly, $H_\mathrm{Y}(\nu)$ and $H_\mathrm{X}(\mu)$ have the first zero-crossings at $\nu = \pm 1/Y$ and $\mu = \pm 1/X$, respectively. Thus, the center lobe around zero, which has a low-pass filtering effect, becomes narrower with increasing sizes X and Y of the filter window. The product $X^2 Y^2 |H_\mathrm{Y}(\nu)|^2 |H_\mathrm{X}(\mu)|^2$ therefore represents the effects of the size and shape that define the window area of the spatial filter.

As may easily be supposed by considering the behavior of $1/\mathrm{sinc}(x)$, the function $|H_\mathrm{c}(\mu)|^2$ of (2.15) periodically diverges with respect to μ, and it is hard to understand the effect of periodicity simply. Alternatively, let us consider a function $|H_\mathrm{Xc}(\mu)|^2$ defined by

$$|H_\mathrm{Xc}(\mu)|^2 = |H_\mathrm{X}(\mu)|^2 |H_\mathrm{c}(\mu)|^2 = \left(\frac{\sin \pi\mu n p}{n \sin \pi\mu p}\right)^2 , \qquad (2.17)$$

where the relation $X = np$ is used. The parameter n in this case represents the number of slits included in length X. The function $|H_\mathrm{Xc}(\mu)|^2$, depicted in Fig. 2.5a, demonstrates periodic peaks at $\mu = 0, \pm 1/p, \pm 2/p, \ldots$, giving selectivity in the spatial frequency domain of μ.

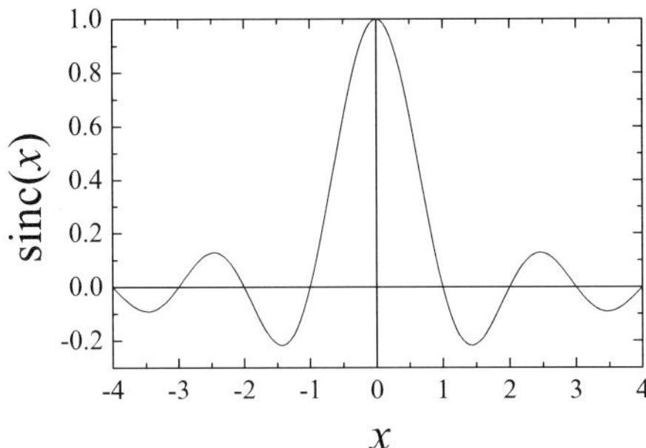

Fig. 2.4. Behavior of a sinc function, $\operatorname{sinc}(x) = \sin \pi x / \pi x$

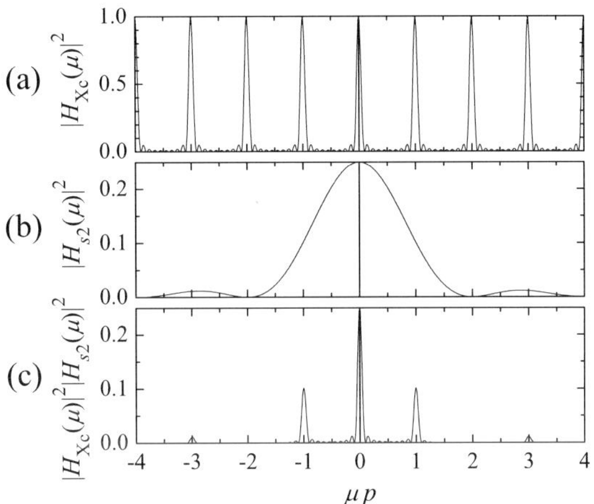

Fig. 2.5. Behaviors of the power spectra: (**a**) $|H_{\mathrm{Xc}}(\mu)|^2$ given by (2.17), (**b**) $|H_{\mathrm{s2}}(\mu)|^2$ given by (2.21) with (2.22), and (**c**) their product $|H_{\mathrm{Xc}}(\mu)|^2|H_{\mathrm{s2}}(\mu)|^2$

The behavior of $|H_{\mathrm{s}}(\mu)|^2$ is estimated by specifying the transmittance function $h(x)$. Here we consider two types of transmittance (**a**) sinusoidal transmittance $h_1(x)$ and (**b**) rectangular transmittance $h_2(x)$, illustrated in Fig. 2.6, which are mathematically given by

$$h_1\left(x\right) = \frac{1}{2}\left(1 + \cos\frac{2\pi}{p}x\right) , \tag{2.18}$$

$$h_2\left(x\right) = \begin{cases} 1, & \left(mp - \dfrac{w}{2}\right) \le x \le \left(mp + \dfrac{w}{2}\right), \\ 0, & \text{otherwise,} \end{cases} \tag{2.19}$$

where w $(0 < w < p)$ denotes the width of one slit, as shown in Fig. 2.3. The power spectra $|H_{\mathrm{s1}}(\mu)|^2$ and $|H_{\mathrm{s2}}(\mu)|^2$ of the transmittance functions $h_1(x)$ and $h_2(x)$ are derived by substituting (2.18) and (2.19) in (2.16) as, respectively [15] (Appendix D),

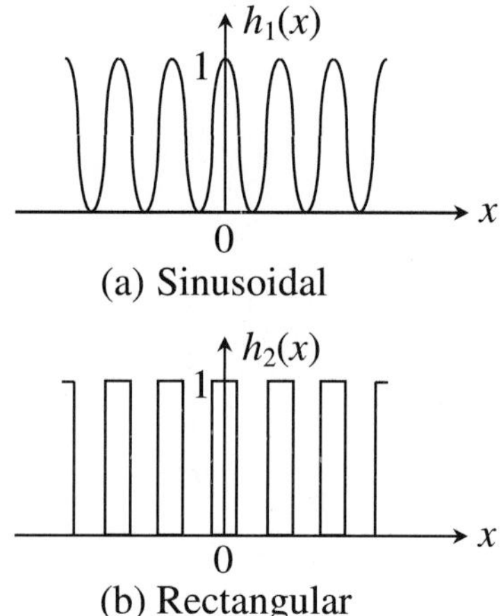

(a) Sinusoidal

(b) Rectangular

Fig. 2.6. Two types of transmittance functions: (**a**) sinusoidal and (**b**) rectangular

$$|H_{s1}(\mu)|^2 = \left(\frac{\sin \pi \mu p}{\pi \mu p}\right)^2 \left[\frac{1}{2(1 - \mu^2 p^2)}\right]^2, \tag{2.20}$$

$$|H_{s2}(\mu)|^2 = \left(\frac{\sin \pi \mu w}{\pi \mu p}\right)^2. \tag{2.21}$$

These spectra are illustrated in Fig. 2.7 and demonstrate low-pass filtering characteristics similar to the function in Fig. 2.4. Generally, rectangular transmittance is easier in manufacturing the spatial filter than sinusoidal transmittance. Comparison of the two spectra in Fig. 2.7 shows no significant difference in their behavior, so rectangular transmittance may be used in normal situations instead of sinusoidal. On the other hand, rectangular transmittance $h_2(x)$ in (2.19) is a function of the slit width w. Figure 2.8 shows the variation of the power spectrum $|H_{s2}(\mu)|^2$ for different values of w. The width

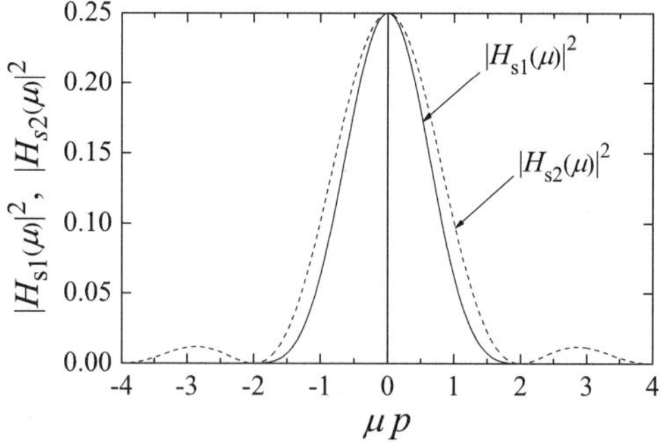

Fig. 2.7. Power spectra $|H_{s1}(\mu)|^2$ and $|H_{s2}(\mu)|^2$ given by (2.20) and (2.21) for the sinusoidal and rectangular transmittance functions $h_1(x)$ and $h_2(x)$, respectively

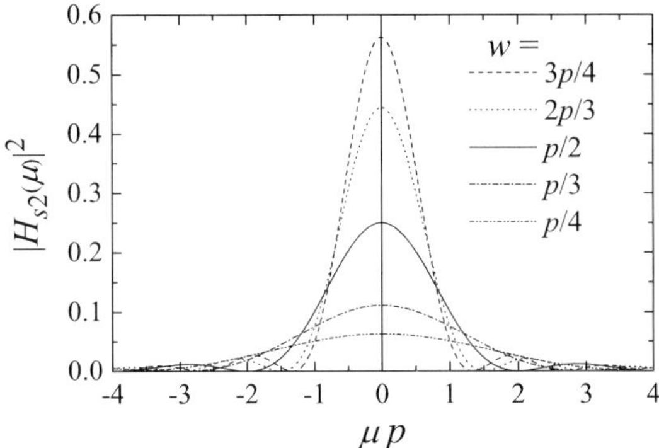

Fig. 2.8. Behavior of the power spectrum $|H_{s2}(\mu)|^2$ for different values of slit width w

of the center lobe and the first zero point become small with increasing slit width w. As will be described in the next section, the spatial filter should be constructed so that the filter will select a spatial frequency component specified by $\mu = 1/p$. This condition requires that the value of the spectrum $|H_{s2}(\mu)|^2$ at $\mu = 1/p$ be as large as possible. The optimum value of w for this requirement is given by [15]

$$w = \frac{p}{2} \, , \tag{2.22}$$

and with this slit width, the spectrum $|H_{s2}(\mu)|^2$ becomes zero at $\mu = 2(m + 1)/p$, $(m = 0, 1, 2, \ldots)$. This behavior is depicted in Fig. 2.5b.

By assuming $XY = 1$ for simplicity, the total power spectrum $|H(\mu, \nu)|^2$ in (2.12) at $\nu = 0$ is then given by the product $|H_{Xc}(\mu)|^2|H_{s2}(\mu)|^2$ for the rectangular-type spatial filter with rectangular transmittance, and illustrated in Fig. 2.5c, where $|H_Y(0)|^2 = 1$. This spectrum demonstrates narrow-band peaks at $\mu = 0$ and $\pm 1/p$. For actually observed temporal signals, the positive frequency domain makes sense in the power spectrum. Thus, the spatial filter selects two spatial frequency components of $\mu = 0$ and $1/p$. The peak at $\mu = 0$ selects a low-frequency component that corresponds to slowly fluctuating bias or pedestal components [5], but they are not useful for spatial filtering velocimetry. Methods for eliminating low-frequency components will be described in Sect. 5.1. The frequency component selected at $\mu = 1/p$ provides periodic signals having a temporal frequency $f = \mu v = v/p$ and, thus, is used for determining the velocity. The power spectrum shown in Fig. 2.5c also contains higher order frequency peaks at $\mu = \pm 3/p, \pm 5/p, \ldots$, but these peaks are substantially attenuated and usually do not contribute to the generation of periodic signals.

In consequence, the periodic transmittance generates narrow-band-pass peaks distributed at intervals of $1/p$ in the power spectrum. The rectangular window having size X produces a low frequency peak at $\mu = 0$. The

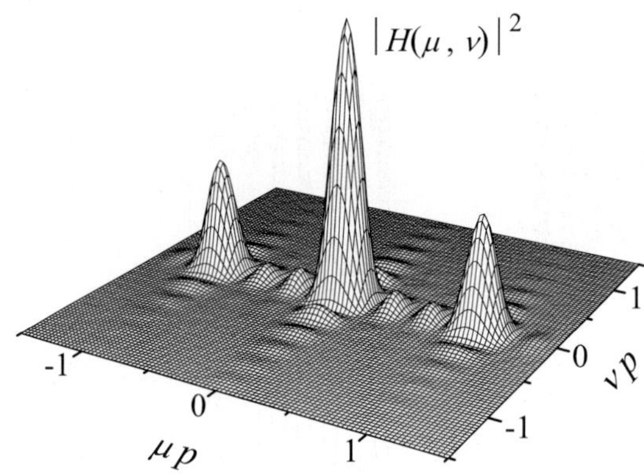

Fig. 2.9. Power spectral distribution $|H(\mu,\nu)|^2$ for a rectangular-type spatial filter with rectangular transmittance

transmittance function within one slit yields a lower frequency passband which covers mainly the range lower than $2/p$ and suppresses or attenuates high-order peaks in the spectrum. On the other hand, the behavior of the power spectrum $|H(\mu,\nu)|^2$ with respect to the spatial frequency axis ν is simply determined by the spectrum $|H_{\mathrm{Y}}(\nu)|^2$ of (2.13), consisting of the sinc function shown in Fig. 2.4. This is an effect of the rectangular window having size Y in the y direction. Finally, a typical illustration of the power spectral distribution $|H(\mu,\nu)|^2$ in axes μ and ν is depicted in Fig. 2.9 for a rectangular-type spatial filter with rectangular transmittance.

2.3 Power Spectra for Typical Spatial Filters

The power spectrum for a rectangular-type spatial filter is given by (2.12) as a product of component spectra which are characterized by each contribution of geometric parameters specifying the spatial filter. In this section, alternative expressions of the power spectrum are given for some typical spatial filters which have different shapes of filter window for sinusoidal or rectangular transmittance. For rectangular transmittance, the condition $w = p/2$ in (2.22) is employed. The expressions obtained here will be used for discussions of spatial filtering characteristics in Sect. 2.4.

Figure 2.10 illustrates transmission-grating spatial filters having (**I**) rectangularly restricted, (**II**) circularly restricted, and (**III**) Gaussian-weighted windows, each for (**a**) sinusoidal and (**b**) rectangular transmittance functions. Transmittance functions and their power spectra for these spatial filters are written below.

(1) *Rectangular type with sinusoidal transmittance* [(**I**) (**a**)]

From (2.18), the transmittance $h(x,y)$ is expressed by

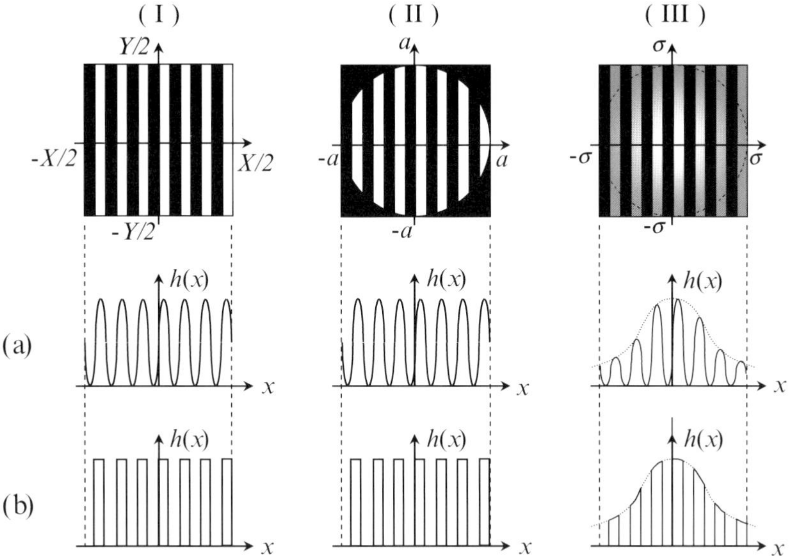

Fig. 2.10. Three different types of spatial filters: (**I**) rectangular, (**II**) circular, and (**III**) Gaussian, each for (**a**) sinusoidal and (**b**) rectangular transmittance functions

$$h\left(x,y\right) = \begin{cases} \dfrac{1}{2}\left(1+\cos\dfrac{2\pi}{p}x\right), & -\dfrac{X}{2} \le x \le \dfrac{X}{2}\,, \ -\dfrac{Y}{2} \le y \le \dfrac{Y}{2}\,, \\ 0\,, & \text{otherwise.} \end{cases} \qquad (2.23)$$

Using (2.23) in (2.5) yields the power spectrum as (Appendix E.1)

$$\left|H\left(\mu,\nu\right)\right|^2 = \frac{X^2 Y^2}{4}\left|H_Y\left(\nu\right)\right|^2 \left\{H_X\left(\mu\right) + \frac{1}{2}\left[H_X^-\left(\mu\right) + H_X^+\left(\mu\right)\right]\right\}^2, \qquad (2.24)$$

where

$$H_X^-\left(\mu\right) = \frac{\sin\pi\left(\mu-\dfrac{1}{p}\right)X}{\pi\left(\mu-\dfrac{1}{p}\right)X}\,,$$

$$H_X^+\left(\mu\right) = \frac{\sin\pi\left(\mu+\dfrac{1}{p}\right)X}{\pi\left(\mu+\dfrac{1}{p}\right)X}\,.$$

(2) *Rectangular type with rectangular transmittance* [(**I**) (**b**)]

From (2.19), the transmittance $h(x,y)$ is given by

$$h\left(x,y\right) = \begin{cases} 1, & \left(mp - \dfrac{w}{2}\right) \le x \le \left(mp + \dfrac{w}{2}\right), \\ 0, & \text{otherwise,} \end{cases} \tag{2.25}$$

for infinite ranges of x and y. This function is expressed by the Fourier series, as

$$h\left(x,y\right) = \frac{1}{2} + \sum_{m=1}^{\infty} \frac{2}{m\pi} \sin \frac{m\pi w}{p} \cos \frac{2m\pi}{p} x \, , \tag{2.26}$$

and if we substitute $w = p/2$ of (2.22) in (2.26), the transmittance is derived as

$$h\left(x,y\right) = \begin{cases} \dfrac{1}{2} + \displaystyle\sum_{m=1}^{\infty} (-1)^{m-1} \dfrac{2}{(2m-1)\pi} \cos\left[\dfrac{2\left(2m-1\right)\pi}{p} x\right], \\ \qquad\qquad -\dfrac{X}{2} \le x \le \dfrac{X}{2} \, , \ -\dfrac{Y}{2} \le y \le \dfrac{Y}{2} \, , \\ 0, \quad \text{otherwise.} \end{cases} \tag{2.27}$$

The second term on the right-hand side in the above equation expresses a sum of cosine functions with different amplitudes and spatial frequencies. For each cosine function, an analogy to the derivation of (1) is applied. Therefore, the power spectrum is obtained as

$$\left|H\left(\mu,\nu\right)\right|^2 = \frac{X^2 Y^2}{4} \left|H_Y\left(\nu\right)\right|^2 \left[H_X\left(\mu\right) + H_{Xm}\left(\mu\right)\right]^2 \, , \tag{2.28}$$

where

$$H_{Xm}\left(\mu\right) = \sum_{m=1}^{\infty} (-1)^{m-1} \frac{2}{(2m-1)\pi} \left[H_{Xm}^{-}\left(\mu\right) + H_{Xm}^{+}\left(\mu\right)\right] \, ,$$

and

$$H_{Xm}^{-}\left(\mu\right) = \frac{\sin \pi \left(\mu - \dfrac{2m-1}{p}\right) X}{\pi \left(\mu - \dfrac{2m-1}{p}\right) X} \, ,$$

$$H_{Xm}^{+}\left(\mu\right) = \frac{\sin \pi \left(\mu + \dfrac{2m-1}{p}\right) X}{\pi \left(\mu + \dfrac{2m-1}{p}\right) X} \, .$$

(3) *Circular type with sinusoidal transmittance* [(**II**) (**a**)]

In this type, the sinusoidal transmittance is restricted by a circular window having radius a, as shown in Fig. 2.10IIa. Thus, the transmittance function is written as

$$h\left(x,y\right) = \begin{cases} \dfrac{1}{2}\left(1 + \cos \dfrac{2\pi}{p} x\right), & x^2 + y^2 \le a^2 \, , \\ 0, & \text{otherwise.} \end{cases} \tag{2.29}$$

By performing the integration of (2.5) with (2.29) over the circular window area, the power spectrum is (Appendix E.2)

$$|H\left(\mu,\nu\right)|^2 = \pi^2 a^4 \left\{ H_{\mathrm{J}}\left(\mu,\nu\right) + \frac{1}{2}\left[H_{\mathrm{J}}^-\left(\mu,\nu\right) + H_{\mathrm{J}}^+\left(\mu,\nu\right)\right]\right\}^2 , \qquad (2.30)$$

where

$$H_{\mathrm{J}}\left(\mu,\nu\right) = \frac{J_1\left(2\pi a\sqrt{\mu^2+\nu^2}\right)}{2\pi a\sqrt{\mu^2+\nu^2}} ,$$

$$H_{\mathrm{J}}^-\left(\mu,\nu\right) = \frac{J_1\left[2\pi a\sqrt{\left(\mu-\dfrac{1}{p}\right)^2+\nu^2}\right]}{2\pi a\sqrt{\left(\mu-\dfrac{1}{p}\right)^2+\nu^2}} ,$$

$$H_{\mathrm{J}}^+\left(\mu,\nu\right) = \frac{J_1\left[2\pi a\sqrt{\left(\mu+\dfrac{1}{p}\right)^2+\nu^2}\right]}{2\pi a\sqrt{\left(\mu+\dfrac{1}{p}\right)^2+\nu^2}} ,$$

where J_1 is a Bessel function of the first order.

(4) *Circular type with rectangular transmittance* [(II) (b)]

From (2.27), the transmittance in this case is written as

$$h\left(x,y\right) = \begin{cases} \dfrac{1}{2} + \displaystyle\sum_{m=1}^{\infty} (-1)^{m-1}\dfrac{2}{(2m-1)\pi}\cos\left[\dfrac{2\left(2m-1\right)\pi}{p}x\right], \\ \qquad\qquad\qquad\qquad\qquad\qquad x^2+y^2 \le a^2 , \\ 0, \quad \text{otherwise.} \end{cases} \qquad (2.31)$$

In the same way as the derivation of (3), the power spectrum is

$$|H\left(\mu,\nu\right)|^2 = \pi^2 a^4 \left[H_{\mathrm{J}}\left(\mu,\nu\right) + H_{\mathrm{Jm}}\left(\mu,\nu\right)\right]^2 , \qquad (2.32)$$

$$H_{\mathrm{Jm}}\left(\mu,\nu\right) = \sum_{m=1}^{\infty}(-1)^{m-1}\frac{2}{(2m-1)\pi}\left[H_{\mathrm{Jm}}^-\left(\mu,\nu\right) + H_{\mathrm{Jm}}^+\left(\mu,\nu\right)\right] ,$$

$$H_{\mathrm{Jm}}^-\left(\mu,\nu\right) = \frac{J_1\left[2\pi a\sqrt{\left(\mu-\dfrac{2m-1}{p}\right)^2+\nu^2}\right]}{2\pi a\sqrt{\left(\mu-\dfrac{2m-1}{p}\right)^2+\nu^2}} ,$$

$$H_{\mathrm{Jm}}^{+}\left(\mu, \nu\right) = \frac{J_1\left[2\pi a\sqrt{\left(\mu + \dfrac{2m-1}{p}\right)^2 + \nu^2}\right]}{2\pi a\sqrt{\left(\mu + \dfrac{2m-1}{p}\right)^2 + \nu^2}} \ .$$

(5) *Gaussian type with sinusoidal transmittance* [(**III**) (**a**)]

The sinusoidal transmittance restricted by the Gaussian-weighted window is written as

$$h\left(x, y\right) = \frac{1}{2}\exp\left(-\frac{x^2 + y^2}{2\sigma^2}\right)\cdot\left(1 + \cos\frac{2\pi}{p}x\right)\ , \tag{2.33}$$

where σ is the effective radius (width $1/e$) of the Gaussian window. Substitution of (2.33) in (2.5) yields the power spectrum (Appendix E.3)

$$\left|H\left(\mu, \nu\right)\right|^2 = \pi^2\sigma^4\left\{H_{\mathrm{G}}\left(\mu, \nu\right) + \frac{1}{2}\left[H_{\mathrm{G}}^{-}\left(\mu, \nu\right) + H_{\mathrm{G}}^{+}\left(\mu, \nu\right)\right]\right\}^2\ , \tag{2.34}$$

where

$$H_{\mathrm{G}}\left(\mu, \nu\right) = \exp\left[-2\pi^2\sigma^2\left(\mu^2 + \nu^2\right)\right]\ ,$$

$$H_{\mathrm{G}}^{-}\left(\mu, \nu\right) = \exp\left\{-2\pi^2\sigma^2\left[\left(\mu - \frac{1}{p}\right)^2 + \nu^2\right]\right\}\ ,$$

$$H_{\mathrm{G}}^{+}\left(\mu, \nu\right) = \exp\left\{-2\pi^2\sigma^2\left[\left(\mu + \frac{1}{p}\right)^2 + \nu^2\right]\right\}\ .$$

(6) *Gaussian type with rectangular transmittance* [(**III**) (**b**)]

Again from (2.27), the rectangular transmittance in the Gaussian type is written as

$$h\left(x, y\right) = \exp\left(-\frac{x^2 + y^2}{2\sigma^2}\right)$$

$$\cdot\left\{\frac{1}{2} + \sum_{m=1}^{\infty}(-1)^{m-1}\frac{2}{(2m-1)\pi}\cos\left[\frac{2(2m-1)\pi}{p}x\right]\right\}\ . \tag{2.35}$$

Thus, the power spectrum is obtained in the same way as the derivation of (2.34) as

$$\left|H\left(\mu, \nu\right)\right|^2 = \pi^2\sigma^4\left[H_{\mathrm{G}}\left(\mu, \nu\right) + H_{\mathrm{Gm}}\left(\mu, \nu\right)\right]^2\ , \tag{2.36}$$

$$H_{\mathrm{Gm}}\left(\mu, \nu\right) = \sum_{m=1}^{\infty}(-1)^{m-1}\frac{2}{(2m-1)\pi}\left[H_{\mathrm{Gm}}^{-}\left(\mu, \nu\right) + H_{\mathrm{Gm}}^{+}\left(\mu, \nu\right)\right]\ ,$$

$$H_{\mathrm{Gm}}^{-}\left(\mu, \nu\right) = \exp\left\{-2\pi^2\sigma^2\left[\left(\mu - \frac{2m-1}{p}\right)^2 + \nu^2\right]\right\}\ ,$$

$$H_{\mathrm{Gm}}^{+}\left(\mu, \nu\right) = \exp\left\{-2\pi^2\sigma^2\left[\left(\mu + \frac{2m-1}{p}\right)^2 + \nu^2\right]\right\}\ .$$

The expression of (2.24) or (2.28) for the power spectrum is different from that of (2.12) including (2.20) or (2.21), respectively, but they are mathematically equivalent to each other. Equations (2.24) and (2.28) may be useful for estimating directly the total distribution of the spectrum $|H(\mu,\nu)|^2$, whereas (2.12) is helpful for understanding each contribution of geometric parameters in the spatial filter to the spectrum. As seen from (2.24), the Fourier spectrum $H(\mu,\nu)$ consists of three frequency components. The first term $H_X(\mu)$ inside the brackets $\{ \ldots \}$ indicates a low-frequency or pedestal component having a peak at $(\mu,\nu) = (0,0)$, and the second term $\left[H_X^-(\mu) + H_X^+(\mu)\right]$ describes desired periodic signal components having peaks at $(\mu,\nu) = (\pm 1/p, 0)$. Thus, the squared absolute of this spectrum or the power spectrum $|H(\mu,\nu)|^2$ is also expected to have three such peaks at the same frequencies. In rectangular transmittance, the Fourier spectrum $H(\mu,\nu)$ in (2.28) also contains higher frequency components having peaks at $(\mu,\nu) = (\pm 3/p, 0), (\pm 5/p, 0), \ldots$ in addition to the three fundamental peaks at $(\mu,\nu) = (0,0)$ and $(\pm 1/p, 0)$. However, the power of those higher order components is low enough to be neglected in comparison with the power of the first-order component at $(\mu,\nu) = (\pm 1/p, 0)$. Note that even-order higher frequency components disappear due to the nature of the spectrum $|H_{s2}(\mu)|^2$ for $w = p/2$ in Fig. 2.8. Although rectangular transmittance is very often employed for actual measurements, sinusoidal transmittance is convenient for theoretical treatments because of its simple function. Actually, a case of $m = 1$ in (2.28) agrees with (2.24) apart from the coefficient for periodic signal terms.

The same discussion as that in the above paragraph can be applied to circular- and Gaussian-type spatial filters. Figure 2.11 illustrates computed examples of the power spectra $|H(\mu,\nu)|^2$ at $\nu = 0$ for (**a**) sinusoidal and (**b**) rectangular transmittance restricted by the three types of windows: rectangular, circular, and Gaussian. The parameter $n = X/p$, which indicates the number of grating lines in the x direction, is set at 5. All of the power spectra show the lower frequency component at $\mu = 0$ and the narrow-band-pass frequency component having a peak at $\mu = 1/p$. The former component is unnecessary for velocity measurements and is usually removed by an electric high-pass filter (HPF). The latter signal component means the desirable selectivity of frequency at $\mu = 1/p$ and, thus, all of the spatial filters in the figure are available for velocity measurements. As to the behaviors of the lower frequency and narrow-band-pass spectra, no significant difference is found between (**a**) the sinusoidal and (**b**) the rectangular transmittance functions. The spectra for the rectangular- and circular-type spatial filters show subsidiary components having quite low power between $\mu = 0$ and $1/p$, but these components hardly affect velocity measurements in normal cases. Since the circular window is generally easy to construct in a range smaller than a few millimeters, the circular-type spatial filter is often used in microscopic measuring systems. The Gaussian-type spatial filter produces no subsidiary component between $\mu = 0$ and $1/p$ and, thus, is ideal. However, it is generally a difficult or complicated task to realize a Gaussian-weighted window.

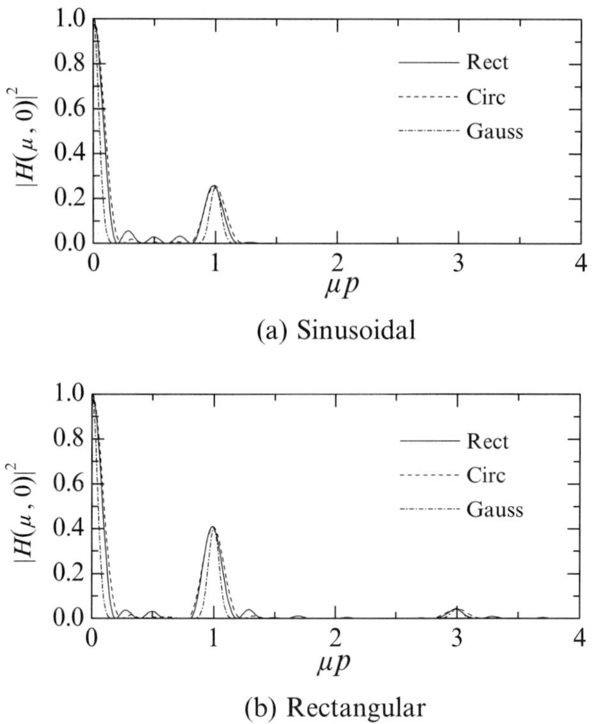

(a) Sinusoidal

(b) Rectangular

Fig. 2.11. Power spectra $|H(\mu,\nu)|^2$ at $\nu = 0$, which are normalized by values at $\mu = \nu = 0$, for (**a**) sinusoidal and, (**b**) rectangular transmittance functions restricted by the three types of windows: rectangular (Rect), circular (Circ), and Gaussian (Gauss)

In comparison with the circular and rectangular types, the Gaussian type is more convenient for mathematical and computational treatments. A possible example of the Gaussian type may be that the probing area, which is defined optically by the circular window of the spatial filter, is illuminated by a Gaussian laser beam of $\mathrm{TEM_{00}}$ mode having a beam width smaller than the probing area.

2.4 Filtering Characteristics

In the spatial filtering method, periodic output signals carrying velocity information are produced by the narrow-band frequency component selected at $\mu = 1/p$ in the power spectrum of the filter's transmittance. The object's velocity can be determined from the central frequency of such periodic signals in the temporal frequency domain. Thus, the quality of output signals is primarily regulated by the filtering characteristics of the narrow-band-pass peak in the power spectrum. Here we investigate the spatial filtering characteristics, particularly the spectral bandwidth and the central frequency of the periodic signal component in the power spectra of spatial filters.

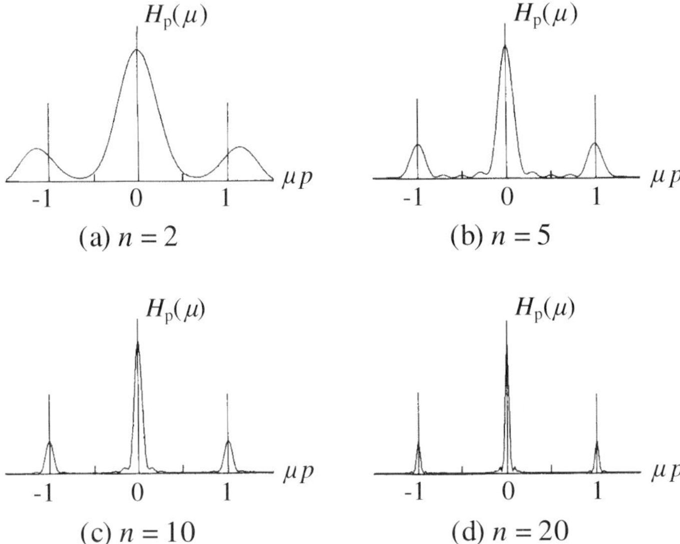

Fig. 2.12. Power spectra $H_{\rm p}(\mu) = |H(\mu)|^2$ for a circular-type grating with sinusoidal transmittance for different numbers of grating lines, n [25]

2.4.1 Spectral Bandwidth

Figure 2.12 [25] shows power spectra $H_{\rm p}(\mu) = |H(\mu)|^2$ for the circular-type grating with sinusoidal transmittance, numerically computed for four different numbers of grating lines, n. The spectrum $H_{\rm p}(\mu)$ is obtained by integrating the spectrum $|H(\mu,\nu)|^2$ with respect to the spatial frequency ν and is more rigorous than $|H(\mu,0)|^2$ for estimating the spectral characteristics $G_{\rm p}(f)$ of output signals obtained from the photodetector. The parameter n is defined by the ratio of the window size (diameter $2a$ for the circular type) to the period p of the grating lines as [25]

$$n = \frac{2a}{p} . \tag{2.37}$$

It is seen from the figure that the spectral bandwidth of the signal component centered at $\mu = 1/p$ becomes small with an increasing number of grating lines, n. To evaluate the spectral bandwidth, we introduce a new parameter D as

$$D = \frac{B}{\mu_0} = p\,B , \tag{2.38}$$

where $\mu_0 = 1/p$ and B are the fundamental spatial frequency given by a reciprocal of the grating line interval p and the half-value full width of the peak at $\mu = \mu_0$, respectively. The former value normally agrees with the central frequency of the signal component. This parameter D denotes the bandwidth of the peak spectrum normalized by the fundamental spatial frequency and is called the *specific bandwidth*. For the circular-type sinusoidal-transmission grating shown in Fig. 2.10IIa, the specific bandwidth D_c is derived approximately as

$$D_\text{c} = \frac{3.233}{n\pi} \qquad \text{(circular)} . \qquad (2.39)$$

The spectral bandwidth is inversely proportional to the number of grating lines, n. The spectral broadening degrades the selectivity of the spatial filter and limits the basic accuracy for measurements of the central frequency in output signals. In addition, the broad peaks for small n make it difficult to remove the lower frequency component with an electric HPF, since the two components centered at $\mu = 0$ and $1/p$ are closely distributed in their tails. Therefore, large number, n, of grating lines is desired generally. In the same way as above, expressions for the specific bandwidth of a rectangular- and Gaussian-type sinusoidal-transmission gratings, shown in Figs. 2.10Ia and IIIa, respectively, are also derived as

$$D_\text{r} = \frac{2.783}{n\pi} , \qquad \text{(rectangular)} , \qquad (2.40)$$

$$D_\text{g} = \frac{1.665}{n\pi} , \qquad \text{(Gaussian)} . \qquad (2.41)$$

Figure 2.13 illustrates the dependence of the specific bandwidth on the number of grating lines, n, given in (2.39)–(2.41) for the three types of sinusoidal-transmission grating spatial filters. With an increasing number of grating lines, n, the specific bandwidth of the signal component decreases, and, then, the selectivity becomes higher. Comparison of the three types indicates that the specific bandwidth for the Gaussian type is smallest for a given number n. Although this difference is due to the window shape of the spatial filter, it may cause no significant defect in the determination of the central frequency. Thus, the difference in the three types will not be an essential factor when we choose the window shape.

Here we consider the basic accuracy for determination of the central frequency on the basis of the specific bandwidth. The central frequency may be assumed to be measurable within an error of $\pm B/2$. Then, the basic accuracy is estimated by the ratio ε of the error $\pm B/2$ to the central frequency μ_0 as

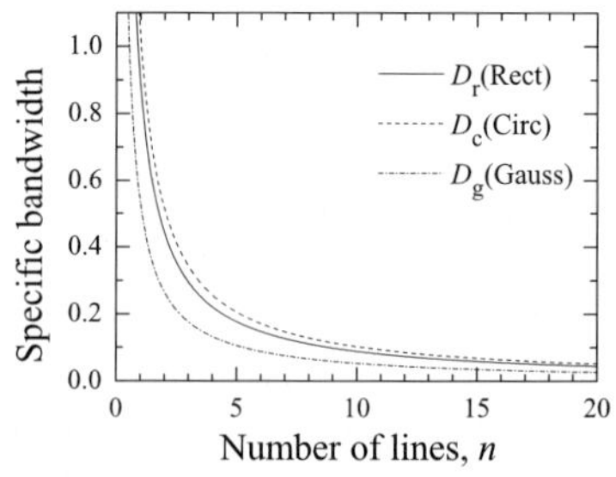

Fig. 2.13. Dependence of the specific bandwidth on the number of grating lines, n

$$\varepsilon = \pm \frac{B}{2\mu_0} = \pm \frac{D}{2} \; . \tag{2.42}$$

It should be noted, however, that the expected accuracy in actual measurements is determined substantially by the ability of the signal-processing system employed.

2.4.2 Central Frequency

As seen from Fig. 2.12, the central frequency of the peak spectrum deviates from $\mu = 1/p$ when $n = 2$. The peak deviation becomes negligibly small for a larger number of grating lines, n. If measuring circumstance or system configuration requires a small number of grating lines, this effect causes significant errors in the measured frequency. The deviation is due to spectral distribution of the pedestal component centered at $\mu = 0$. For a small number n of grating lines, the spectral bandwidth of the pedestal component also becomes broad, and its tail is significantly overlapped by the signal component around $\mu = 1/p$. Figure 2.14 illustrates the power spectra for a circular-type sinusoidal-transmission grating with $n = 2$, 2.6, and 3. In graphs (**a**) and (**c**), the central frequency of the signal component deviates from $\mu = 1/p$ to the higher and lower frequency regions, respectively, whereas in (**b**), the central frequency almost agrees with $\mu = 1/p$ despite the small number n. This difference in the behavior of the central frequency is due to the form of superposition of the pedestal and signal components. In the lower three graphs (**a'**), (**b'**), and (**c'**) of Fig. 2.14, the signal component is calculated and plotted separately from the pedestal component (shown by the *broken curve*), and the central frequency does not deviate at all. For a large number n of grating lines, the tail of the pedestal component sufficiently attenuates in the frequency region of the signal component and has no influence on it.

Here we consider a deviation error in percent defined by the frequency deviation $\Delta\mu$ from the fundamental spatial frequency $\mu = \mu_0 = 1/p$, normalized by μ_0 as

$$\mathrm{Error} = \frac{\Delta\mu}{\mu_0} \times 100 \; . \tag{2.43}$$

Figure 2.15 demonstrates the deviation error evaluated from computed power spectra for a circular-type sinusoidal transmission grating as a function of the number of grating lines, n. The error is large for small n, especially for $n < 5$, and becomes small with increasing number n. From this evaluation, the number n of grating lines larger than 10 is recommended to suppress the deviation error to less than 1%.

The Gaussian-type transmission grating is free from this problem. The pedestal component given by the Gaussian function in (2.34) rapidly attenuates as the spatial frequency μ increases and no subsidiary oscillation appears in its spectral distribution. Thus, the tail of the pedestal component does not affect the signal component. By its nature, the Gaussian type is considered

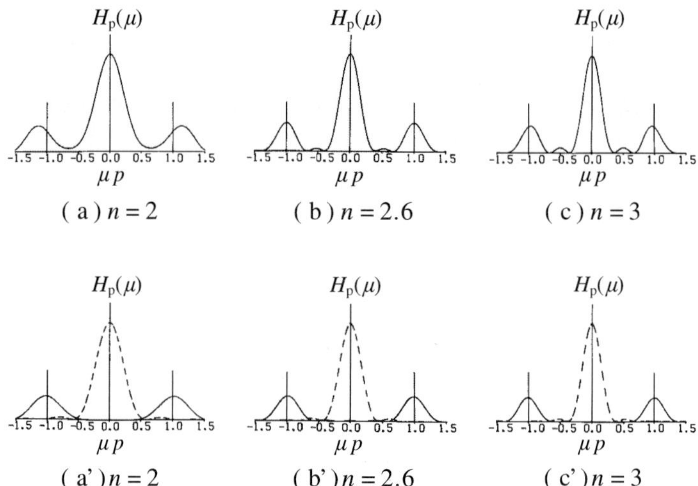

Fig. 2.14. Power spectra for a circular-type sinusoidal-transmission grating with $n = 2$, 2.6, and 3. In the lower three graphs, the signal (*solid curve*) and pedestal (*broken curve*) components are separately calculated and plotted

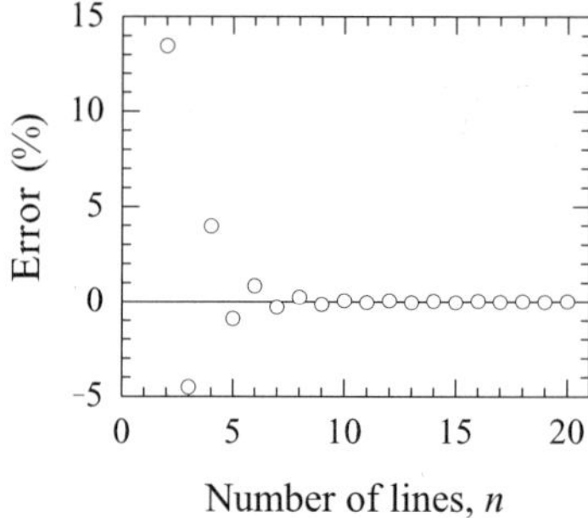

Fig. 2.15. Deviation error of the central frequency from $\mu = 1/p$ as a function of the number of grating lines, n [25]

ideal in spatial filtering characteristics. On the other hand, the rectangular type as well as the circular type suffer from the central frequency deviation for smaller n. Since the deviation is due to the presence of the pedestal tail component, it should be removed. Generally, the pedestal component can be eliminated by an electric HPF. Note, however, that, even if the pedestal component is removed sufficiently in such a way, the electric HPF is unable to eliminate the tail component of pedestal that is embedded in the signal component and to avoid deviation of the central frequency. To suppress the effect, a specific optical method may be employed to eliminate the pedestal component before signal processing. For example, the differential detection method is useful for this purpose and will be described in Sect. 5.1.2.

2.4.3 Direction of Grating Lines

Up to this subsection, the grating lines of the spatial filter have been assumed to be orthogonal to the direction of the object's movement. Next we investigate the effect of change in the moving direction, with respect to the grating lines, on spatial filtering characteristics. Figure 2.16 illustrates the object moving in the x' direction at an angle θ to the x axis which is perpendicular to the grating lines. By rotating the coordinate axes given by

$$\begin{cases} x' = x\cos\theta + y\sin\theta\,, \\ y' = -x\sin\theta + y\cos\theta\,, \end{cases} \tag{2.44}$$

the power spectrum $H'_{\mathrm{p}}(\mu,\nu)$ for a circular-type sinusoidal-transmission grating is written as [25] (Appendix F)

$$|H(\mu,\nu)|^2 = \pi^2 a^4 \left\{ H_{\mathrm{J}}(\mu,\nu) + \frac{1}{2}\left[H_{\mathrm{J}\theta}^-(\mu,\nu) + H_{\mathrm{J}\theta}^+(\mu,\nu)\right] \right\}^2 , \tag{2.45}$$

where $H_{\mathrm{J}}(\mu,\nu)$ is identical to that in (2.30) and

$$H_{\mathrm{J}\theta}^-(\mu,\nu) = \frac{J_1\left[2\pi a\sqrt{\left(\mu - \dfrac{\cos\theta}{p}\right)^2 + \left(\nu - \dfrac{\sin\theta}{p}\right)^2}\right]}{2\pi a\sqrt{\left(\mu - \dfrac{\cos\theta}{p}\right)^2 + \left(\nu - \dfrac{\sin\theta}{p}\right)^2}}\,,$$

$$H_{\mathrm{J}\theta}^+(\mu,\nu) = \frac{J_1\left[2\pi a\sqrt{\left(\mu + \dfrac{\cos\theta}{p}\right)^2 + \left(\nu + \dfrac{\sin\theta}{p}\right)^2}\right]}{2\pi a\sqrt{\left(\mu + \dfrac{\cos\theta}{p}\right)^2 + \left(\nu + \dfrac{\sin\theta}{p}\right)^2}}\,.$$

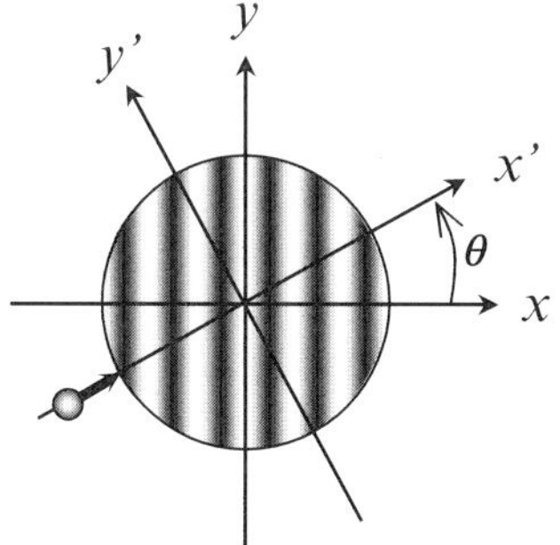

Fig. 2.16. Case of the object movement in the x' direction at an angle θ to the x axis which is perpendicular to the grating lines

This equation indicates that the narrow band-pass spectrum of the signal component appears at $(\mu, \nu) = (\pm \cos\theta/p, \pm \sin\theta/p)$. The central frequency f_0' of this spectrum in the temporal frequency domain is, then, given from (1.1) and (2.10) by

$$f_0' = \frac{M v_0}{p} \cos\theta = f_0 \cos\theta \,, \tag{2.46}$$

and the measured velocity in this case is given by

$$v_0' = \frac{p}{M} f_0' = \frac{p}{M} f_0 \cos\theta = v_0 \cos\theta \,. \tag{2.47}$$

The two equations above express that a substantial decrease in the factor $\cos\theta$ affects the values of the central frequency and the velocity being measured. As long as the angle θ is unknown, the real velocity v_0 cannot be derived. The measured value is the cosine component of v_0 in the x direction. Figure 2.17a [25] shows computed power spectra $H_p'(\mu)$ for four different angles of θ in the use of the circular-type sinusoidal-transmission grating. These spectra are obtained by integrating $H_p'(\mu, \nu)$ with respect to ν. The number of grating lines, n, is set at 4. A deviation error of the central frequency from $\mu = 1/p$ is plotted in Fig. 2.17b. With increasing angle θ, the central frequency becomes lower making the error larger and nearer the pedestal component. Since the pedestal component in the spectra is not influenced by a change in angle θ, it keeps the consistent spectral form. For a larger angle θ, thus, the signal and pedestal component spectra overlap significantly. This creates difficulty in the perfect removal of the pedestal component by using the electric HPF without affecting the signal spectral component. To overcome this problem, the pedestal component should be eliminated by the optical system after all.

The decrease of the central frequency in the signal component spectrum means a substantial broadening of the spectral width. For a given central

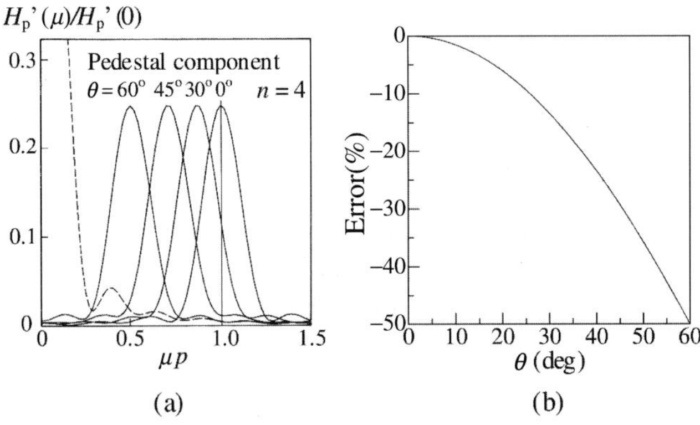

(a)   (b)

Fig. 2.17. (a) Power spectra $H_p'(\mu)$ for four different angles θ using the circular-type sinusoidal-transmission grating [25], and (b) the deviation error of the central frequency from $\mu = 1/p$

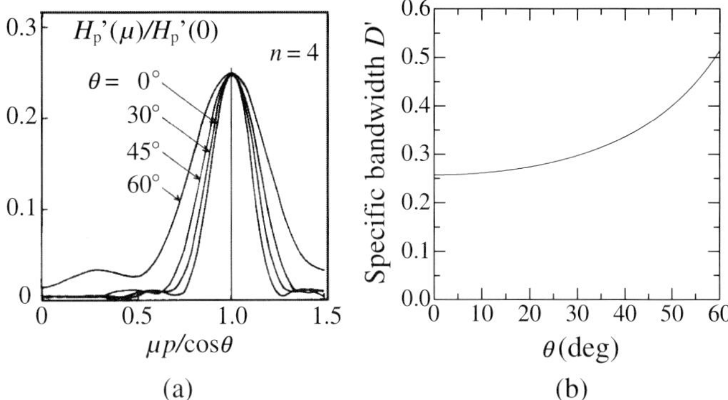

Fig. 2.18. (a)Spectrum broadening of the signal component for four different angles of θ and (b)the specific bandwidth as a function of the angle

frequency $\mu'_0 = (1/p)\cos\theta$, the specific bandwidth in this case is written as

$$D' = \frac{B}{\mu'_0} = \frac{pB}{\cos\theta} = \frac{D}{\cos\theta} . \tag{2.48}$$

Figures 2.18a and b demonstrate the spectrum broadening of the signal component and the specific bandwidth as a function of angle θ. In graph (**a**), the pedestal component is omitted, and the spatial frequency axis is normalized with the deviated central frequency $\mu'_0 = (1/p)\cos\theta$. Thus, the bandwidth in this graph shows directly the specific bandwidth given by (2.48). An increase in angle θ broadens the bandwidth and degrades the selectivity of the spatial filter. The directional effect described above can be applied to measurements of two-dimensional components of the object velocity, which will be described in Sect. 5.1.5.

2.5 Parameters of the Spatial Filter

To realize required filtering characteristics in a given measurement circumstance, the physical parameters of the spatial filter should be appropriately designed. This section provides some useful considerations for determining the parameters that characterize the performance of spatial filtering velocimetry. According to the characteristics discussed in the above sections, the required spatial filter may be specified by considering the following parameters [56] :

1. transmittance function
2. filter window
3. interval of grating lines
4. number of grating lines.

These parameters are described in the following subsections.

2.5.1 Transmittance Function

Section 2.2 described two typical transmittance functions: sinusoidal and rectangular. Investigation of the power spectra in Sect. 2.3 reveals that the sinusoidal transmittance is better for filtering properties because it does not generate higher order signal frequency components and gives mathematically simple expressions. However, the exact realization of the sinusoidal transmittance for a spatial filtering device is not easy and often costly. Most usual measurements are satisfied by rectangular transmittance, though it generates unwelcome higher order signal frequency components. These components have negligibly low intensity and no significant effect on the fundamental frequency component, as long as the number of grating lines, n, is about 10 or more. Rectangular transmittance can easily be realized by photoprinting or photoetching techniques and is also commercially available (Ronchi ruling, for example). Measurements in microscopic regions often require a small number n, with which the signal spectrum is broadened, and, thus, rectangular transmittance may be problematic since the tails of neighboring spectral peaks possibly overlap. Roughly for $n < 4$, sinusoidal transmittance should be used. It may be possible to design other functions such as a triangular function for transmittance. Though they are not general, they are used for alternative spatial filtering devices, providing that the generation of higher order signal components is carefully considered.

2.5.2 Filter Window

The filter window is an aperture which is usually attached to the transmission grating to define the detecting area with a certain shape and size. Then, the window defines the cross-sectional area of the probing volume by means of optical imaging. The total power of light received by the photodetector is determined by this window. However, the main consideration should be given to the number of grating lines, n. With a given size X for the window in the x direction, the number n is determined by X/p, in which p is the grating line interval. On one hand, ensuring a large number n requires a large window. On the other hand, high spatial resolution in velocity measurements requires a small window. The window size is, thus, determined by a compromise between the number, n, of grating lines and the spatial resolution.

The three representative shapes of the window are described in Fig. 2.10; rectangular, circular, and Gaussian windows. Probably, the circular or rectangular aperture is a general choice. These shapes are very easily designed and constructed, and their properties are also sufficiently practicable as long as the number n of grating lines is relatively large. The circular aperture is particularly favorable if a small window of a few millimeters or less is required, since a small circular aperture or pinhole is more easily fabricated than a rectangular one. The Gaussian aperture provides us with the best filtering quality and mathematical simplicity, but it is not a simple task to realize an actual Gaussian filtering device. Thus, the Gaussian window may be suitable for theoretical studies. Other shapes can also be used if there is a specific requirement.

2.5.3 Intervals of Grating Lines

Usually, the intervals of grating lines may be appropriately selected from some discrete values offered by manufacturers. The selection should be made by considering the available number of grating lines for a given window size. The upper limit of the interval is given by the minimum permissible number of grating lines. The lower limit is determined by the image size of an object such as a particle. From an estimation of the size effect which will be discussed in Sect. 2.6, we recommend that the interval be larger than twice the image size. In making the transmission grating, a line interval larger than hundred microns is easily fabricated without particular difficulties, but an extremely small interval may need advanced manufacturing techniques.

2.5.4 Number of Grating Lines

The number of grating lines, n, is the most important parameter of the spatial filter that characterizes the performance of spatial filtering velocimetry. From the investigation of filtering characteristics on the basis of power spectra, the number n should be designed as large as possible to ensure high selectivity. In Fig. 2.13, the specific bandwidth decreases with an increase in the number n and the decreasing slope becomes nearly flat for $n > 10$. This property indicates that further improvement in selectivity is not largely expected for $n > 20$. The minimum permissible number of grating lines may be evaluated from the basic accuracy ε given by (2.42). To ensure $\varepsilon \leq \pm 5\%$, the specific bandwidth D_c for the circular-type of sinusoidal-transmission grating should be ≤ 0.1, and this requires $n > 10$. Equation (2.42) often yields an overestimation for restricting the available number n. With empirical knowledge [25], even $n = 5$–10 can be used successfully for general purposes.

A typical example of a spatial filter may be provided by rectangular transmittance restricted by a circular or rectangular window having the number of grating lines, $n = 10$–20. The geometric sizes of the window and grating line interval are, then, decided by considering the mean size of images of scattering objects, the probing cross-sectional area, and the optical magnification.

2.6 Effects of Scattering Objects

As described in Sect. 2.1, the spatial power spectrum of output signals, $G_\mathrm{p}(\mu, \nu)$, is characterized by the power spectrum $H_\mathrm{p}(\mu, \nu)$ for a spatial filter, assuming that the power spectrum $F_\mathrm{p}(\mu, \nu)$ for the scattering object is almost uniform in the spatial frequency range up to around $\mu = \pm 1/p$ (see

Fig. 2.2). This assumption holds when the mean size of images of scattering objects such as particles is sufficiently small in comparison with the interval p of the grating lines. Unless this is satisfied, properties of the spectrum $F_{\rm p}(\mu, \nu)$ may affect the output spectrum $G_{\rm p}(\mu, \nu)$. In this section, the effects of the spectrum $F_{\rm p}(\mu, \nu)$ on output signal characteristics are investigated.

2.6.1 Deviation of the Central Frequency

Let us first treat scattering particles as objects and consider the effect of their mean size. A circular particle image having a diameter $2b$ is assumed to move on a circular-type sinusoidal-transmission grating having interval p. Figure 2.19a shows one-dimensional computed power spectra $H_{\rm p}(\mu)$ for a spatial filter with the number of grating lines, $n = 4$, and $F_{\rm p}(\mu)$ for a particle image having three different sizes, which are specified by the ratio of the image diameter to the grating line interval, $2b/p$. As the particle image is comparable in size with the grating interval p such as $2b/p = 0.8$ in Fig. 2.19a, the power spectrum $F_{\rm p}(\mu)$ for the particle image is no longer uniform in the spatial frequency domain and attenuates considerably with increasing spatial frequency μ. With an increase in the ratio $2b/p$ from 0.2 to 0.8, this attenuation becomes noticeable. Thus, at the central frequency, $\mu = 1/p$, of a narrow-band-pass spatial filter, the power spectrum $F_{\rm p}(\mu)$ for the larger particle image is more reduced in power. Since the power spectrum $G_{\rm p}(\mu)$ of output signals is given by the product of $H_{\rm p}(\mu)$ and $F_{\rm p}(\mu)$, $G_{\rm p}(\mu)$ is affected by the decreasing characteristics of $F_{\rm p}(\mu)$, as shown in Fig. 2.19b. In this figure, the power spectrum $G_{\rm p}(\mu)$ for a larger ratio of $2b/p$ becomes lower, and the resultant central frequency deviates toward the lower frequency. The significant slope of the power spectrum $F_{\rm p}(\mu)$ around $\mu = 1/p$ results in the central frequency deviation. If $F_{\rm p}(\mu)$ is nearly uniform around

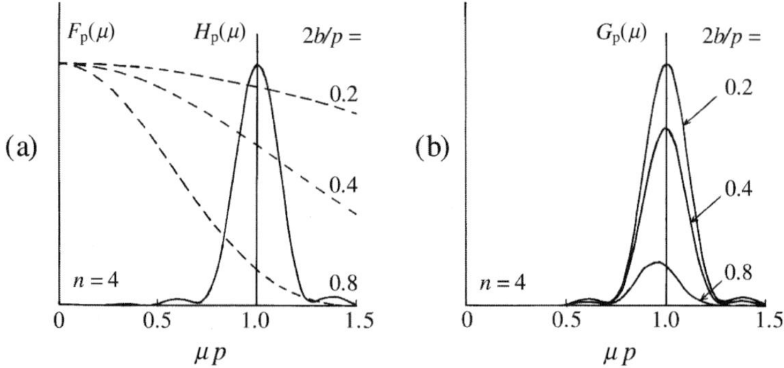

Fig. 2.19. (a) Power spectra $H_{\rm p}(\mu)$ for a spatial filter with the number of grating lines, $n = 4$, and $F_{\rm p}(\mu)$ for the particle image having three different sizes $2b$ relative to the grating line interval p. (b) The corresponding power spectrum $G_{\rm p}(\mu)$ obtained by the product of $H_{\rm p}(\mu)$ and $F_{\rm p}(\mu)$

$\mu = 1/p$, no deviation occurs, even if $F_{\mathrm{p}}(\mu)$ is low in power. This deviation may cause significant errors in the measured central frequency and, thus, velocity.

To treat the frequency deviation quantitatively, a one-dimensional intensity distribution $f(x)$ of a circular particle image is assumed to take a Gaussian form as

$$f(x) = \frac{1}{\sqrt{2\pi}b_{\mathrm{g}}}\exp\left(-\frac{x^2}{2b_{\mathrm{g}}^2}\right) , \tag{2.49}$$

where b_{g} is the standard deviation of $f(x)$, corresponding to the effective radius of the particle image. Then, the power spectrum $F_{\mathrm{p}}(\mu)$ of the particle image is derived by using (2.6) as

$$F_{\mathrm{p}}(\mu) = \exp\left(-4\pi^2 b_{\mathrm{g}}^2 \mu^2\right) . \tag{2.50}$$

For mathematical simplicity, the Gaussian-type sinusoidal-transmission grating shown in Fig. 2.10IIIa is introduced here instead of the circular type. A one-dimensional power spectrum $G_{\mathrm{p}}(\mu)$ of output signals is derived by using (2.34) and (2.50). The result is given in Appendix G. The central frequency, giving a peak of the signal component, is derived in this case as

$$\mu = \frac{1}{p} \cdot \frac{n^2}{\left(\dfrac{2b_{\mathrm{g}}}{p}\right)^2 + n^2} . \tag{2.51}$$

If the particle image is sufficiently small in comparison with the grating interval p or the value of $2b_{\mathrm{g}}/p$ is negligibly small compared to the value of n, then the central frequency agrees nearly with $\mu = 1/p$. Figure 2.20 demonstrates the dependence of the central frequency normalized with $1/p$ for the ratio $2b_{\mathrm{g}}/p$. An increase in the ratio $2b_{\mathrm{g}}/p$ makes the normalized central frequency μp lower. This tendency is quite remarkable for a small number of grating lines, n. The frequency deviation due to the particle size should be taken into account when the number n of grating lines is less than 10. A similar effect due to scattering particle size has also been reported in studies of laser Doppler velocimetry [57, 58].

2.6.2 Visibility of Output Signals

Another significant effect due to scattering particle size, which is demonstrated in Fig. 2.19b, is the reduction in power of the signal component in comparison with the pedestal component. This reduction means a decrease in the contrast of the periodic signal amplitude (ac component) to its average (dc component) and, thus, degrades the signal quality. A signal from a small particle image is fully modulated by the grating, as illustrated in Fig. 2.21a, whereas a large particle image may give a fractionally modulated signal, as in Fig. 2.21b. The large particle image extends over plural opaque bars and slits of the grating

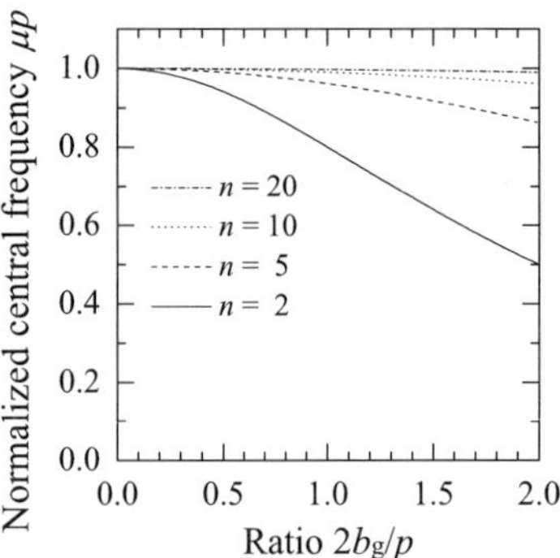

Fig. 2.20. Dependence of the normalized central frequency on the ratio of the particle image diameter to the grating line interval, $2b_\mathrm{g}/p$

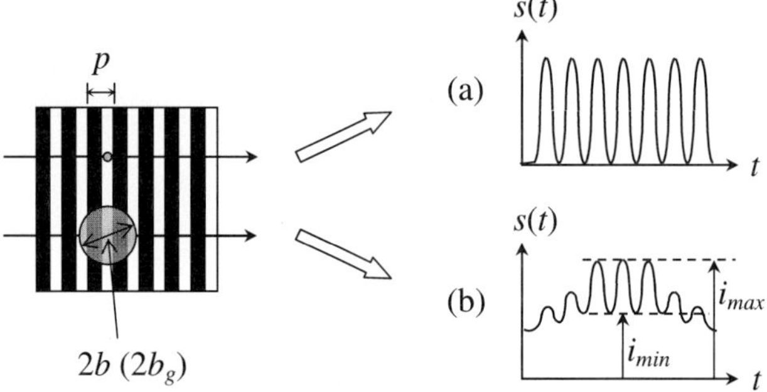

Fig. 2.21. Small and large particle images passing on the transmission grating and their resultant signals (a) and (b), respectively

and, thus, averages a variation of the transmitted light intensity. The modulation of signals produced by the image passage on the grating is, then, substantially reduced in Fig. 2.21b.

Figure 2.22 shows numerically simulated output signals obtained from circular images having different diameters passing on a Gaussian-type rectangular-transmission grating. Signals (**a**)–(**f**) are plotted for different ratios, $2b/p$, of the particle image diameter to the grating line interval. Signals (**a**) and (**b**) for smaller particle images are fully modulated. The depth of the modulation is reduced with an increasing ratio of $2b/p$, and signal (**e**) is almost averaged out. The degree of signal modulation may be evaluated by "*visibility*" V that is defined as [5]

$$V = \frac{i_{\max} - i_{\min}}{i_{\max} + i_{\min}} = \frac{\text{ac amplitude}}{\text{dc component}}, \tag{2.52}$$

where $i_{\max}$ and $i_{\min}$ are the maximum and minimum levels of the periodic signal amplitude, as shown in Fig. 2.21b. In the following simplified cases,

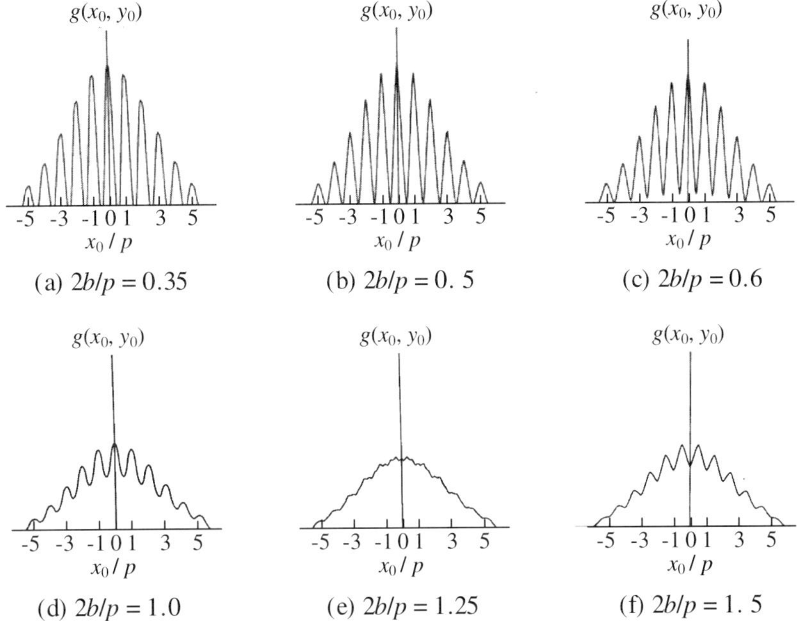

Fig. 2.22. Numerically simulated output signals from circular images having different diameters $2b$ passing on the Gaussian-type rectangular-transmission grating with the line interval p

expressions for visibility are derived by specifying the intensity distribution function $f(x, y)$ of a single particle image and the grating transmittance $h(x, y)$.

(1) A circular image of diameter $2b$ passing on a sinusoidal-transmittance filter

The two functions $f(x, y)$ and $h(x, y)$ are written as

$$f(x, y) = \begin{cases} I_0, & x^2 + y^2 \leq b^2, \\ 0, & \text{otherwise,} \end{cases} \tag{2.53}$$

$$h(x, y) = \frac{1}{2}\left(1 + \cos\frac{2\pi}{p}x\right) . \tag{2.54}$$

Substitution of these two functions in (2.1) yields the following result for the modulated intensity of transmitted light(Appendix H.1):

$$g(t) = \frac{I_0 \pi b^2}{2}\left[1 + A_1 \cos\left(\frac{2\pi v_x}{p}t\right)\right] , \tag{2.55}$$

$$A_1 = \frac{2J_1\left(\dfrac{2\pi b}{p}\right)}{\dfrac{2\pi b}{p}} ,$$

where v_x is the x component of the image velocity and J_1 is a Bessel function of the first order. If the light intensity received by the detector is assumed to be proportional to $g(t)$, visibility V_1 is given by

$$V_1(b) = |A_1| \ .$$
(2.56)

In the same way above, the visibility may be calculated for other cases below. For the following, then, only the resultant equations are given here.

(2) A circular image of diameter $2b$ passing on a rectangular-transmittance filter:

$$g(t) = \frac{I_0 \pi b^2}{2} \left[1 + 2 \sum_{m=1}^{\infty} (-1)^{m-1} \frac{4}{(2m-1)\pi} A_2 C_m(t) \right] \ ,$$
(2.57)

$$A_2 = \frac{J_1\left(\dfrac{2\,[2m-1]\,\pi b}{p}\right)}{\dfrac{2\,[2m-1]\,\pi b}{p}} \ ,$$

$$C_m(t) = \cos\left[\frac{2(2m-1)\pi v_x}{p} t\right] \ ,$$

$$V_2(b) = \left| \sum_{m=1}^{\infty} (-1)^{m-1} \frac{8}{(2m-1)\pi} A_2 \right| \ .$$
(2.58)

(3) An image having a Gaussian intensity profile with a standard deviation b_g, passing on a sinusoidal-transmittance filter (Appendix H.2):

$$g(t) = I_0 \pi b_g^2 \left[1 + A_3 \cos\left(\frac{2\pi v_x}{p} t\right) \right] \ ,$$
(2.59)

$$V_3(b_g) = A_3 = \exp\left(-\frac{2\pi^2 b_g^2}{p^2}\right) \ .$$
(2.60)

(4) An image having a Gaussian intensity profile with a standard deviation b_g, passing on a rectangular-transmittance filter:

$$g(t) = I_0 \pi b_g^2 \left[1 + \sum_{m=1}^{\infty} (-1)^{m-1} \frac{4}{(2m-1)\pi} A_4 C_m(t) \right] \ ,$$
(2.61)

$$A_4 = \exp\left[-\frac{2(2m-1)^2 \pi^2 b_g^2}{p^2}\right] \ ,$$

$$V_4(b_g) = \left| \sum_{m=1}^{\infty} (-1)^{m-1} \frac{4}{(2m-1)\pi} A_4 \right| \ .$$
(2.62)

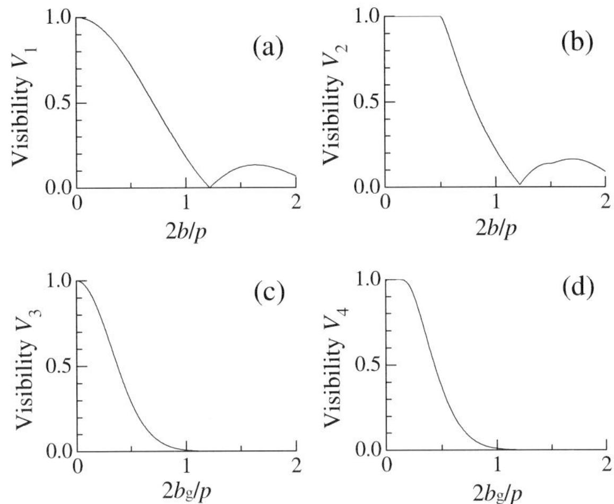

Fig. 2.23. Visibility of output signals as a function of the ratio, $2b/p$ or $2b_{\mathrm{g}}/p$, of the particle image diameter to the interval of grating lines

The case for the rectangular-transmittance filter may be more realistic than the sinusoidal one. A model of the Gaussian-shaped image is sometimes of practical interest since it may correspond approximately to the defocused image and to the edge-blurred image that is due to the point spread of an optical imaging system, as will be described in Chap. 3.

Figure 2.23 shows plots of visibility V as a function of the ratio $2b/p$ or $2b_{\mathrm{g}}/p$, of the particle image diameter to the interval of grating lines. Higher visibility is obtained for the ratio $2b/p$ or $2b_{\mathrm{g}}/p$ equal to and less than about 0.5–0.6. With $2b_{\mathrm{g}}/p = 0.5$, the error in the frequency deviation is estimated by (2.51) or Fig. 2.20 at about 6% for $n = 2$, which is substantially the minimum number of grating lines in the principle of spatial filtering velocimetry. Thus, the effects of the particle image size may be negligible for $2b/p$ or $2b_{\mathrm{g}}/p \leq 0.5$–0.6. Note that the amplitude of output signals $g(t)$ is proportional to the value of $\pi b^2 V$ or $\pi b_{\mathrm{g}}^2 V$. This means that extremely small particles may decrease the light intensity received by the detector, even if the visibility is nearly unity. The small amplitude of output signals degrades the signal-to-noise ratio in the presence of electrical noises. Thus, the signal quality depends on both visibility and the absolute signal amplitude as long as particle size is concerned. The value of $\pi b^2 V_1$ becomes maximum for $2\pi b/p = 1.841$ and, then, the optimum ratio of particle image diameter to grating line interval, $2b/p$, is 0.586 in this simplified case [5]. This is also permissible with respect to the above condition of $2b/p \leq 0.5$–0.6 for suppressing the frequency deviation and producing high visibility.

It may be interesting to note that the above problem of size effects is very similar to laser Doppler velocimetry [59,60], in which visibility is described as a function of the ratio of the scattering particle diameter to the fringe spacing.

Spatial filtering velocimetry treats visibility as a function of the particle image diameter for a given line interval. Then, visibility should also be an important factor when the optical imaging system accompanies a focusing error. A defocused image makes its effective size larger and reduces visibility. Therefore, visibility is one of the factors that limit the focusing depth in the probe volume. The optical imaging system and the probe volume will be described in Chap. 3. Measurements of visibility can be used for determining particle sizes and detecting of the focusing position. These topics will be discussed in Sect. 6.5.

2.6.3 Light Scattering by Spherical Particles

In the above discussion, the particle image was assumed to have a circular cross-section or a two-dimensional Gaussian profile. This model is, however, sometimes not applicable to spherical scattering particles. The light scattering by a very large sphere compared to the wavelength of illuminating light may be treated by a geometric optics approximation [61]. Figure 2.24 illustrates the imaging of a spherical particle under the illumination of a collimated beam. The light scattered in the near forward direction is imaged onto the plane of a transmission grating by a lens. According to the geometric optics approximation, scattered light is expressed by the sum of reflected, refracted, and diffracted rays. Under conditions of dark-field illumination, diffracted rays are not used for imaging because the directions of illumination and detection are clearly different. Higher order refracted or internally

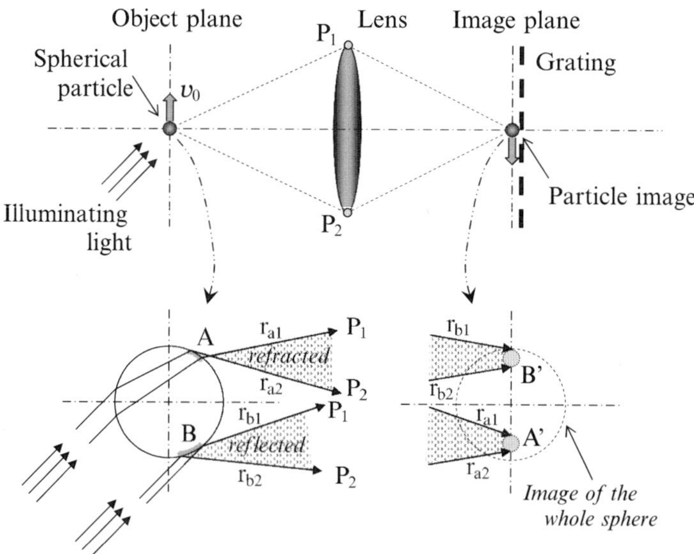

Fig. 2.24. Imaging of a spherical particle illuminated by a collimated beam

reflected rays are rather attenuated and their contribution is usually negligible. Thus, reflected and twice-refracted rays are considered here. The direction of reflected and refracted rays follows Snell's law at the particle–medium interface.

In Fig. 2.24, two refracted rays emerging from the sphere and directing toward the upper and lower edges $\mathbf{P}_1$ and $\mathbf{P}_2$ of the lens are denoted by $\mathbf{r}_{a1}$ and $\mathbf{r}_{a2}$, respectively. In the same way, two reflected rays with the same condition are $\mathbf{r}_{b1}$ and $\mathbf{r}_{b2}$. In this illustration, the refractive index of the particle to medium is assumed to be larger than unity. The hatched region between the two rays $\mathbf{r}_{a1}$ and $\mathbf{r}_{a2}$ represents the flux of refracted rays that is received by the lens and contributes to the imaging. Since this light flux emerges from a surface fraction $\mathbf{A}$ of the sphere, its image $\mathbf{A}'$ becomes a cross section of $\mathbf{A}$ projected on the image plane. The flux of reflected rays defined by the region between $\mathbf{r}_{b1}$ and $\mathbf{r}_{b2}$ forms the image of surface fraction $\mathbf{B}$, denoted by $\mathbf{B}'$ in the image plane. In usual measurement circumstances, the scattering of ambient light by a sphere contributes to forming a circular cross-sectional image of the whole sphere. Because the ambient light is usually diffused and not collimated, emerging rays to be imaged cover the whole cross section. This image is, however, considerably weak in intensity unless sufficient diffuse illumination is actively introduced. The images $\mathbf{A}'$ and $\mathbf{B}'$ of surface fractions may be observed as bright spots. Therefore, output signals are produced mainly by the passage of two bright spots on the grating rather than that of the whole sphere image. In this case, the effects of particle size may be estimated more properly for the bright spot images of the surface fractions. Roughly speaking, the size of the bright spot image may be 0.1–0.2 of the whole sphere-image diameter, though the situation depends on the scattering condition and the imaging optical system. In consequence, there is the possibility of extending the size range of measurable spheres by spatial filtering velocimetry. However, the above phenomenon should be carefully considered when the measurement of visibility is applied to the determination of spherical particle sizes.

2.7 Requirements for Scattering Objects

The technique of spatial filtering velocimetry depends on the spatial distribution of the image intensity of the light scattered or radiated from the moving object being measured. The suitability of the image intensity distribution for spatial filtering velocimetry can be estimated by the spatial power spectrum of the distribution. As already shown in Fig. 2.2, the spatial filter selects the narrow band spectral component centered at the spatial frequency $\mu = \mu_0 = 1/p$ from the power spectrum $F_\mathrm{p}(\mu, \nu)$ of the image intensity distribution. Thus, $F_\mathrm{p}(\mu, \nu)$ must contain a substantial level of power around $\mu = 1/p$. Figures 2.25a and b show illustrations of possible power spectral

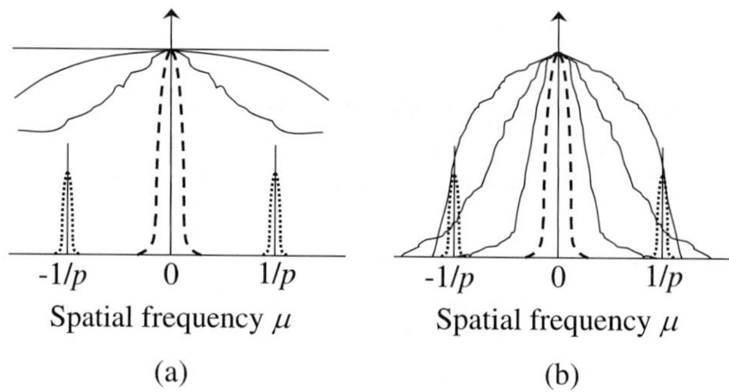

Fig. 2.25. Possible distribution forms of the power spectrum $F_\mathrm{p}(\mu)$: (**a**) suitable and (**b**) unsuitable for spatial filtering velocimetry. The dotted spectrum shows $H_\mathrm{p}(\mu)$ for the spatial filter

examples $F_\mathrm{p}(\mu)$ suitable and unsuitable for spatial filtering velocimetry, respectively. The dotted spectrum shows $H_\mathrm{p}(\mu)$ for the spatial filter. Spectral distributions in figure (**a**) have a comparable level of power at $\mu = 1/p$ with the dc component at $\mu = 0$, and the signal frequency spectrum selected by the spatial filter may keep a measurable level of power. Some spectra in figure (**b**) have lower power at $\mu = 1/p$ in comparison with the dc component, $F_\mathrm{p}(0)$. Periodic signals in this case are of low contrast (or visibility), and the signal-to-noise ratio (SNR) may be poor. Other spectra in figure (**b**) show a dramatic change in power around $\mu = 1/p$, which causes the deviation error in the resultant central frequency, as described in Sect. 2.6.1. In consequence, the requirement for scattering objects is that the level of the spectrum $F_\mathrm{p}(\mu, \nu)$ around $\mu = 1/p$ should be larger than a quarter to half of the dc component $F_\mathrm{p}(0,0)$ and should also be comparably uniform. Three typical examples of objects are treated for this subject in the following subsections.

2.7.1 Small Particles

In spatial filtering velocimetry, measurements in fluids require small suspended particles to scatter the illuminating or natural light for imaging. The upper limit in particle size is given by estimations of the frequency deviation and the visibility lowering, as discussed in Sects. 2.6.1 and 2.6.2. The useful criterion is that the diameter of particle images on the grating should be smaller than about half of the grating line interval p. Particles smaller than this may ensure successful measurements, but too small particles scatter insufficient intensity of light for an acceptable signal-to-noise ratio in the background of the ambient light. A recommendation from the practical point of view is that the ratio of particle image diameter to grating interval, $2b/p$, should be in and around a range of 0.05–0.5. The shape and material of particles are less important factors for SFV measurements as long as particle images are obtained with an acceptable contrast to the background. Images with insufficient contrast yield an unsuitable power spectrum, as illustrated in Fig. 2.25b.

One of the factors influencing SFV measurements is the particle concentration. In Sect. 2.6, output signals from a single particle image crossing the grating lines in a plane were investigated. The situation treated here is that many particles are present together in the probing volume. We assume that all particles are moving with the same velocity in the same direction. Since particles are distributed randomly in the fluid, the phases of periodic intensity variations produced by the passing of their images on the grating must be random. This leads to considerable mutual cancellation. The output intensity signal resulting from the nth particle in the probing volume may be written as

$$g\left(t\right) = a_{\mathrm{n}} \cos\left(2\pi f_0 t + \varphi_{\mathrm{n}}\right) , \tag{2.63}$$

where a_{n} and φ_{n} indicate the amplitude and phase of the intensity signal from the nth particle, respectively, and f_0 denotes the central frequency that is proportional to v_{x}/p. The values of φ_{n} are considered random and not correlated with a_{n}, and, thus, averaging over a time long enough to include contributions of many different particles may result in [5]

$$\overline{\left[g\left(t\right)\right]^2} = \frac{1}{2} \sum_n \overline{a_{\mathrm{n}}^2} . \tag{2.64}$$

If N identical particles are sampled for averaging, this simple case yields

$$\overline{\left[g\left(t\right)\right]^2} = N a^2 , \tag{2.65}$$

where a indicates a root mean intensity from the contribution of each particle. Thus, the total root-mean-square intensity of signals, which is obtained by the square root of (2.65), is proportional to the square root of the number of particles, $\sqrt{N}$, in the probing volume. As seen from the expressions for the intensity signal $g(t)$ derived in Sect. 2.6.2, the level of the dc component produced by a single particle is proportional to the cross-sectional area of the particle. Then, dc levels due to many particles are directly additive, except for extremely high concentration. Therefore, the ratio of a periodic signal (ac component) to the mean detector output (dc component) decreases with increasing number N of particles, and the signal quality deteriorates in a high concentration of particles.

As the number of particles increases to infinity, the signal-to-noise ratio approaches zero. If an extremely large number of small particles are imaged on the grating plane, such a projected scene looks like a uniform cloud of particles. In this case, the movement of the cloud is expected to yield no periodic variation in the intensity of light transmitted through the grating because the slits of the grating always contain the same number of particle images statistically. The power spectrum of a uniform cloud of particle images may have a dominant pedestal component whose tail attenuates dramatically as the frequency increases, such as one of the examples shown in Fig. 2.25b. The problem of signal quality for many particles is almost the same as that for laser Doppler velocimetry [5]. In LDV, however, the signal does not completely disappear at high concentration since it is due to coherent signals from the optical beating of light scattered from different particles. There is clearly no possibility of using coherent signals in spatial filtering velocimetry. The effect of particle number density on the signal-to-noise ratio was investigated by Chan and Ballik [44] in a comparison study between the differential-type laser Doppler velocimeter and the white light fringe image velocimeter. The joint probability density of the amplitude and phase of SFV signals was studied by Michel and his group [62] to increase the measurement accuracy when many particles pass through the probe volume.

2.7.2 Rough Surfaces

If a rough surface illuminated with the white light is imaged and a randomly varying light intensity distribution is observed on the grating, the velocity of the surface may be measured with spatial filtering velocimetry. The suitability of the surface for SFV measurements depends on the contrast and mean lateral fluctuation of a random pattern on the surface image. A low-contrast image decreases the ratio of the signal amplitude to the dc component and degrades the signal quality. The mean lateral fluctuation of the random intensity pattern is estimated by the correlation length τ_c. For simplicity, we consider one-dimensional image intensity distribution $f(x)$ in the spatial axis x on a grating. Let the fluctuation of image intensity from its mean level be expressed by

$$\Delta f(x) = f(x) - \langle f(x) \rangle \, , \tag{2.66}$$

where $< \cdots >$ stands for an ensemble average. The autocorrelation function $\gamma(\tau_x)$ is given by

$$\gamma(\tau_x) = \int_{-\infty}^{\infty} \Delta f(x) \, \Delta f(x + \tau_x) \, \mathrm{d}x \, . \tag{2.67}$$

In usual cases, the function $\gamma(\tau_x)$ decreases as τ_x increases. The correlation length τ_c is defined by the shifting length τ_x that gives $1/e$ or half of the maximum correlation value $\gamma(0)$. Since the length τ_c gives the mean lateral extent of a correlated area in the random image, $2\tau_c$ can be conveniently regarded as the equivalent particle diameter. In consequence, the condition for the suitability of randomness for SFV measurements may be expressed as

$$0.05 \leq \frac{2\tau_c}{p} \leq 0.5 \tag{2.68}$$

from the discussion of size effects in Sect. 2.6.

If the value of $2\tau_c$ is larger than half of the grating line interval p, the random fluctuation of image intensity is reduced with the lower spatial frequency, and the signal quality is degraded. An extremely small value of τ_c means that the random fluctuation is not resolvable by observation or detection and the resultant signal may lose the periodic intensity variation. The situation is, then, nearly the same as that for a uniform cloud of small particles. There may be some resolutions for this problem. One is the use of images magnified appropriately and another is the use of a speckle pattern illuminated by coherent light or a laser. The latter possibility is described in the next subsection. Cracks and sharp edges may also be possible objects for SFV measurements since they contain higher spatial frequency components. For these objects, however, attention should be given to errors from a sudden change of the signal intensity in the signal processing system.

In relation to rough surfaces, various natural scenes are also considered suitable objects to be measured by the SFV technique. Images of the sea, sky, and land surfaces including forests and roads may be a kind of random radiation intensity distribution. The spatial power spectra of those images, which depend also on the observation optical system, generally contain spatial frequency components to some extent. Then, the central frequency $\mu = 1/p$ of the spatial filter being used can be placed in the frequency range of images, as illustrated in (**a**) of Fig. 2.25. As a matter of fact, the spatial filtering method has been studied since its beginning for sensing the velocity of an airplane with respect to the ground surface by using terrain images [12, 13]. As for the signal quality, the same discussion for the contrast and lateral correlation length of images as rough surfaces can be applied to these natural scenes.

2.7.3 Speckle Pattern

A moving diffuse object illuminated by coherent light produces a dynamic speckle pattern in the diffraction and image fields. Although there are two typical speckle motions of "translation" and "boiling" [7], translational speckles can be used for velocity measurements based on the spatial filtering method [63]. Boiling speckles are ineffective for SFV measurements because of the nature of their motion. Since a speckle pattern has a random intensity distribution, its translation on a grating produces a periodic variation in the transmitted light intensity as well as the image of rough surfaces discussed in the previous subsection. In speckle theory, the lateral correlation length and contrast of the speckle intensity variation are referred to as "speckle size" and "speckle contrast," respectively. By evaluating these values, the quality of SFV signals for a speckle pattern may be discussed in the same way as that for a rough surface. Note that speckle contrast depends on the surface roughness of the diffuse object [64]. Fully developed speckles result in a contrast of almost unity and are good for the spatial filtering method. A low-contrast speckle pattern leads to a decrease in the signal-to-noise ratio.

Usually, the spatial filtering method is applied to moving images or moving intensity patterns in the image plane. For speckles, however, translation in the diffraction field can also be detected by the spatial filtering method. In this case, the relation between the object's velocity and the speckle velocity must be known for the optical system being used. A motion of speckles having both translation and boiling simultaneously may degrade output signals. Since one bright speckle grain may disappear before crossing all grating lines within the entire spatial filter area, the effective number of grating lines that experience translation of the speckle grain substantially decreases. This causes broadening of the signal frequency spectrum and, thus, reduces measurement accuracy. A detailed study [65] has been theoretically and experimentally carried out for the relation between spectral broadening and speckle motion. For an application of the spatial filtering method to a speckle pattern, the optical system for producing speckles should be considered carefully to ensure pure translational motion.

3

Optical System

Since output signals in the SFV method are generated by an image moving on a spatial filter, the image quality may have direct effects on signal quality and, thus, on the performance of SFV systems. Imaging is one of the fundamental optical phenomena and is well described in books on optics [66]. It is probably true that the image formation used in SFV can be referred to the general imaging theory in such books. Therefore, no specific study has been reported on this subject in research on SFV so far. In fact, many optical instruments such as the microscope and telescope are provided with high-quality imaging systems that have been well designed for their purposes, and they may be used for the optical system of SFV in some specified applications. A wide range of applications, however, requires better understanding of the basic relation between image formation and signal quality.

Optical systems used in the SFV method may be divided into illuminating and imaging parts. This chapter is devoted to some basic issues of optical systems which should be considered to obtain clear or sharp, high-contrast, and high-fidelity images. In Sect. 3.1, the resolution of imaging systems is discussed by using the point spread given by the Airy disk and also the transfer function. Section 3.2 discusses the effects of lens aberrations on image performance. Definitions for the depth of focus and the probe volume are next described in Sect. 3.3 under the paraxial approximation. In Sect. 3.4, illumination is briefly considered and, finally, some additional matters for imaging are presented in Sect. 3.5.

3.1 Resolution of Imaging Systems

If imaging is treated only by light rays that lie near the optical axis or are restricted to the framework of paraxial approximation in geometric optics, a point object is imaged to a point. However, a real optical system is accompanied by diffraction effects which form spread images and, thus, two closely spaced point objects are often unresolved in their images. They are fractionally or mostly superposed and the image quality is reduced, which may influence SFV measurements. Though the criterion for the two-point resolution or the so-called Rayleigh criterion can be used as a quality factor for optical systems [55], the transfer function is more practical and informative in various applications.

3.1.1 Point Spread

Generally, imaging is performed by a lens having a circular aperture, and the image of an object is influenced by diffraction due to the aperture. The Fraunhofer diffraction pattern of a circular aperture under plane-wave illumination is given by the two-dimensional Bessel function, which is mathematically proportional to the Fourier transform of the circular transmittance function. The diffraction pattern consists of a central bright spot known as the Airy disk and surrounding rings with rather weak intensity. By this effect, the image of a distant point source or small particle does not appear as a point, but as the above-mentioned diffraction pattern, even if the optical system is perfectly free from lens aberrations. Thus, this pattern represents the impulse response of an ideal (aberration-free) lens and is called a point spread function. The radius of the Airy disk is known as the *Rayleigh criterion*, which gives the minimum separation in the image plane by which two incoherent point sources are "barely resolved." The Rayleigh criterion has traditionally been used to estimate the resolution of imaging systems. Figure 3.1a illustrates the Fraunhofer diffraction of a plane wave with a wavelength λ by a circular lens having a diameter (or a pupil diameter) D and a focal length f. In this figure, the radius r_{a} of the Airy disk is given by

$$r_{\mathrm{a}} = \frac{1.22\lambda f}{D} \,. \tag{3.1}$$

For the surrounding medium with a refractive index of $n \neq 1$, the corresponding wavelength $\lambda_{\mathrm{n}} = \lambda/n$ may be used instead of λ. A small value of the diameter D yields a large value for radius r_{a} and reduces the imaging resolution. This can be a cause of blurred or low-contrast images and the resultant signals are deteriorated.

The diameter of the Airy disk provides the smallest particle image that can be obtained for a given imaging configuration. This diameter d_{a} is given by using (3.1) as

$$d_{\mathrm{a}} = 2r_{\mathrm{a}} = \frac{2.44\lambda f}{D} = 2.44\lambda F \,, \tag{3.2}$$

where $F = f/D$ is the F-number. The above expression corresponds to the imaging of a sufficiently distant point object. In the usual imaging as shown in Fig. 3.1b, the point spread varies with a imaging magnification $M = l_2/l_1$, where l_1 and l_2 denote the distances between the object plane and the lens plane (assumed to be a thin lens) and between the lens plane and the imaging

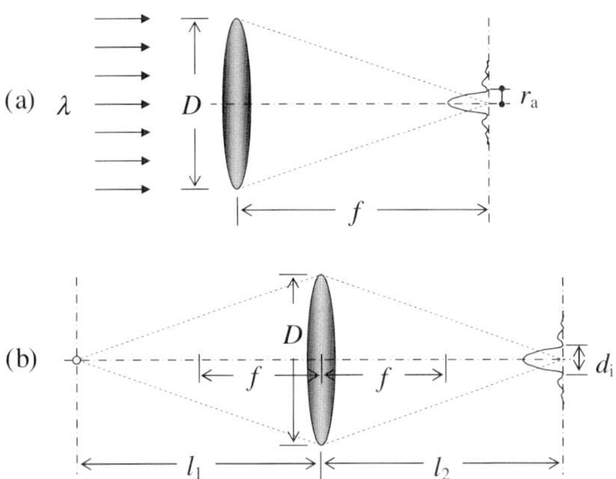

Fig. 3.1. (a) Fraunhofer diffraction by a circular lens and (b) point spread in a single lens imaging system

plane, respectively. The diameter d_i of the Airy disk then becomes a function of the distance l_2 and is given, by taking account of the well-known imaging equation:

$$\frac{1}{l_1} + \frac{1}{l_2} = \frac{1}{f} \, , \tag{3.3}$$

as

$$d_\mathrm{i} = \frac{2.44\lambda l_2}{D} = \frac{2.44\lambda f}{D} \left(1 + M\right) = 2.44\lambda F_\mathrm{e} \, , \tag{3.4}$$

where F_e is the effective F-number given

$$F_\mathrm{e} = \frac{l_2}{D} = \frac{f\left(1 + M\right)}{D} = F\left(1 + M\right) \, . \tag{3.5}$$

Typical calculated values for the diameter d_i of the Airy disk are given in Table 3.1 with the wavelength of $\lambda = 500\,\mathrm{nm}$ and magnifications of $M = 0.5$, 1, and 2. A large F-number (or a small lens aperture for a given focal length f) and a large magnification generate a large point spread. However, note that for a large magnification, the relative size of the point spread to the geometric image is reduced because the geometric image itself is also magnified by M. Table 3.1 shows that the influence of the point spread becomes significant for imaging small particles having a diameter of a few micrometers with small magnifications.

As will be described in Sect. 3.1.2, the imaging of finite-size particles is given by the convolution of the geometric particle image with the point spread function. This means that, as the particle becomes large, geometric imaging becomes dominant and the effect of diffraction may be negligible. When assuming aberration-free imaging and the point spread in the form given by the Bessel function, the image diameter $2b$ of a large particle can be estimated by the following useful formula [3, 67]:

$$2b = \sqrt{(Md_{\mathrm{p}})^2 + d_{\mathrm{i}}^{\,2}}\,, \tag{3.6}$$

where d_{p} is the diameter of a scattering particle. Table 3.2 shows typical values of the image diameter for three finite-size particles, calculated with three different magnifications $M = 1$, 5, and 10, the F-number $F = 2.8$, and a wavelength $\lambda = 500\,\mathrm{nm}$. The difference between the geometric diameter Md_{p} and the estimated diameter $2b$ from (3.6) becomes insignificant with increases in magnification and particle diameter.

If imaging is influenced by diffraction, the size of expected images is larger than the geometric size and, in the worst case, the increased size of particle images results in a decrease of signal visibility and deviation of the central frequency, which were discussed in Sect. 2.6. A small lens aperture may also decrease the power of scattered light received at a detector and reduce the ratio of signal to background noise. Therefore, a smaller F-number is recommended for SFV measurements of small particles in an order of micrometers. This requires a larger lens aperture for a given focal length. Unfortunately, a very large lens aperture generally increases aberrations since paraxial approximation is not applied. However, it is not simple to evaluate quantitatively the total effects of lens aberrations and, then, design of the F-number may be made primarily by estimating the point spread in theory. When the maximum permissible point spread or the diameter d_{i} of the Airy disk is given, a criterion for the F-number can be determined by (3.4) with known wavelength λ and necessary magnification M. For example, the maximum permissible point

Table 3.1. Calculated values for the diameter d_{i} of the Airy disk with $\lambda = 500\,\mathrm{nm}$

F	d_{i} (μm)		
	$M = 0.5$	$M = 1$	$M = 2$
2	3.7	4.9	7.3
2.8	5.1	6.8	10.2
4	7.3	9.8	14.6
5.6	10.2	13.7	20.5
8	14.6	19.5	29.3

Table 3.2. Values of the image diameter $2b$ calculated from (3.6) for three different-size particles with $F = 2.8$ and $\lambda = 500\,\mathrm{nm}$

M	$2b$ (μm)		
	$d_{\mathrm{p}} = 50\,\mu$m	$d_{\mathrm{p}} = 100\,\mu$m	$d_{\mathrm{p}} = 200\,\mu$m
1	50.5	100.2	200.1
5	250.8	500.4	1000.2
10	501.4	1000.7	2000.3

spread of $\sim 10\,\mu\mathrm{m}$ in diameter requires an F-number $F = 4$ for $\lambda = 500\,\mathrm{nm}$ and $M = 1$ in Table 3.1. If some allowance is further desired for a safe design for unexpected sources of image deterioration, the above-determined F-number may be reduced by $1/\sqrt{2}$.

In the above discussion, the effects of point spread were treated in the image plane. If an estimation of resolution is required for the object plane, the following equation may be useful:

$$d_{\mathrm{o}} = \frac{d_{\mathrm{i}}}{M} = \frac{1.22\lambda l_1}{D} = 1.22\lambda F\left(\frac{1}{M} + 1\right) , \tag{3.7}$$

where d_{o} means the minimum diameter of a resolvable circular area in the object plane.

There is a useful approximated expression for the point spread function of a circular aperture, although it is originally given by the Bessel function. The approximated point spread function may be given by a normalized Gaussian distribution defined as [3]

$$p(x) = \exp\left(-\frac{x^2}{2\sigma^2}\right) , \tag{3.8}$$

where

$$\sigma = \frac{\sqrt{2}\,\lambda f}{\pi D}(1 + M) . \tag{3.9}$$

Equation (3.8) provides quite simple mathematical treatments in comparison with the Bessel function, and is often convenient for practical use.

The Rayleigh criterion and point spread effects based on the Airy disk are mainly used for estimation of the imaging on and near the optical axis and, for example, are effective for microscope objectives, which form the image of an object in a relatively small field around the optical axis. If a larger field must be covered, the estimation should be made for the entire area of the image. For this purpose, the following transfer function is more useful and reliable than the estimation using the Airy disk.

3.1.2 Transfer Function

The point spread of optical imaging systems not only increases the diameter of particle images but also reduces the ontrast of images. The previous discussion reveals that the point spread is a serious problem for small particles. As may easily be presumed from this fact, a spatially fine structure or details in an object, which contain higher frequencies, may be more blurred in its image, leading to a significant loss of image contrast. This property can be estimated by a transfer function of the imaging system as a function of the spatial frequency.

Let us briefly review the physical meaning of the transfer function. For simple treatments, the description will be given by one-dimensional functions.

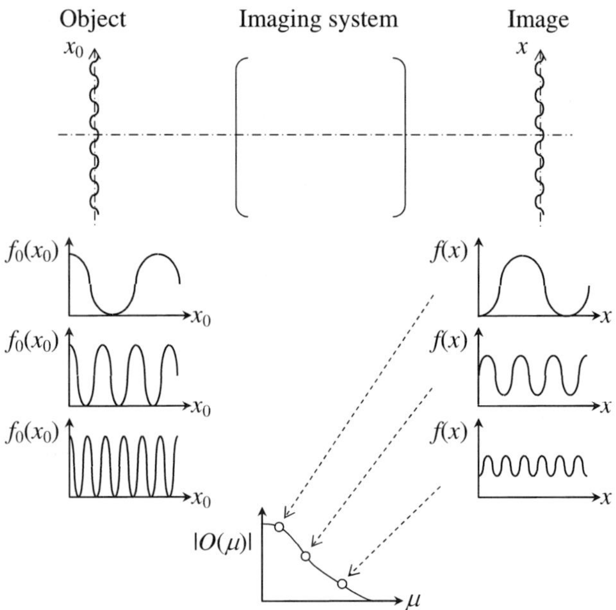

Fig. 3.2. Schematic description of the concept of the transfer function in an optical imaging system

In Fig. 3.2, if an object is assumed to have a spatially sinusoidal intensity distribution $f_0(x_0)$ defined by

$$f_0(x_0) = a_0 + b_0 \cos(2\pi\mu_0 x_0) \qquad (a_0 \geq b_0)$$
$$= a_0 \left[1 + m_0 \cos(2\pi\mu_0 x_0)\right] , \tag{3.10}$$

the corresponding image intensity distribution $f(x)$ may be written as

$$f(x) = a + b \cos(2\pi\mu x - \phi) \qquad (a \geq b)$$
$$= a \left[1 + m \cos(2\pi\mu x - \phi)\right] , \tag{3.11}$$

where μ_0 and μ denote the spatial frequency in the object and image planes, respectively, and m_0 and m are modulation indexes of the object and image intensity distributions, respectively, which are given by

$$m_0 = \frac{b_0}{a_0} \qquad (0 \leq m_0 \leq 1) \tag{3.12}$$

$$m = \frac{b}{a} \qquad (0 \leq m \leq 1) . \tag{3.13}$$

These indexes represent a contrast of the intensity variation. By this imaging, the modulation index m_0 for an object is transferred to m for the image. Thus, a ratio of m to m_0:

$$M\left(\mu\right) = \frac{m}{m_0} \tag{3.14}$$

indicates the quality in the transfer of modulation, which may vary with a change in the spatial frequency, as shown in Fig. 3.2. Then, $M(\mu)$ is generally referred to as a modulation transfer function (MTF). The phase shift ϕ is also a function of the spatial frequency, and $\phi(\mu)$ is referred to as a phase transfer function (PTF) as well. With $M(\mu)$ and $\phi(\mu)$, the function $O(\mu)$ is defined as follows:

$$O\left(\mu\right) = M\left(\mu\right)\exp[\mathrm{i}\phi\left(\mu\right)]\,, \tag{3.15}$$

which is commonly called an optical transfer function (OTF). In SFV measurements, the MTF is an important factor for estimation of imaging properties since it characterizes the reproduction of the contrast of images. The PTF means an in-plane displacement of the intensity distribution as a function of the spatial frequency. This effect may result in phase distortion of images and, thus, the PTF is considered secondarily important. Of course, the MTF with $M(\mu) = 1$ is most desirable in any spatial frequency. Note that $M(\mu) = |O(\mu)|$ in (3.15).

Generally, optical imaging systems can be classified into any of two types with incoherent illumination or coherent illumination. Here, we do not discuss partially coherent illumination, which is beyond the scope of this book. The transfer function is, thus, defined separately for the two imaging systems. For the incoherent imaging system, the transfer function is given in an intensity form, whereas it is treated in an amplitude form for the coherent imaging system. In an incoherent imaging system, the image intensity distribution $f(x, y)$ is given by the convolution of the object intensity distribution $f_0(x_0, y_0)$ with the point spread function $t(x, y)$ of the system, written as

$$f\left(x, y\right) = \int_{-\infty}^{\infty} f_0\left(x_0, y_0\right) t\left(x - x_0, y - y_0\right) \mathrm{d}x_0\,\mathrm{d}y_0\,. \tag{3.16}$$

By Fourier transforming both sides of this equation, we obtain

$$F\left(\mu, \nu\right) = F_0\left(\mu, \nu\right) T\left(\mu, \nu\right)\,, \tag{3.17}$$

where $F(\mu, \nu)$, $F_0(\mu, \nu)$, and $T(\mu, \nu)$ are the Fourier transforms of $f(x, y)$, $f_0(x, y)$, and $t(x, y)$, respectively, and represent the intensity spectra. Equation (3.17) indicates that the input function $F_0(\mu, \nu)$ and the output function $F(\mu, \nu)$ are related in a linear system. The function $T(\mu, \nu)$ operates as the linear filter and is exactly an optical transfer function of the incoherent system. It is commonly known that the OTF for an incoherent system agrees with the autocorrelation of the pupil function of the lens, which includes an amplitude term and also a phase term. For a single thin-lens system as the simplest case, the pupil function is given by the aperture transmittance function of the lens. Thus, the OTF is always unity at zero frequency, and the MTF (or an absolute of the OTF) at any frequency is always less than unity [55].

In a coherent imaging system, the image amplitude distribution $f^{\mathrm{a}}(x,y)$ is given by the convolution of the object amplitude distribution $f_0^{\mathrm{a}}(x_0,y_0)$ with the amplitude point spread function $t^{\mathrm{a}}(x,y)$ of the system and is expressed as

$$f^{\mathrm{a}}(x,y) = \int_{-\infty}^{\infty} f_0^{\mathrm{a}}(x_0,y_0)\, t^{\mathrm{a}}(x-x_0, y-y_0)\, \mathrm{d}x_0\, \mathrm{d}y_0 \; . \tag{3.18}$$

This is the same form of expression as (3.16). Thus, the Fourier transform of the above equation yields

$$F^{\mathrm{a}}(\mu,\nu) = F_0^{\mathrm{a}}(\mu,\nu)\, T^{\mathrm{a}}(\mu,\nu) \; , \tag{3.19}$$

where $F^{\mathrm{a}}(\mu,\nu)$, $F_0^{\mathrm{a}}(\mu,\nu)$, and $T^{\mathrm{a}}(\mu,\nu)$ denote the Fourier transforms of $f^{\mathrm{a}}(x,y)$, $f_0^{\mathrm{a}}(x,y)$, and $t^{\mathrm{a}}(x,y)$, respectively. Equation (3.19) indicates again a relation of the linear system for input $F_0^{\mathrm{a}}(\mu,\nu)$ and output $F^{\mathrm{a}}(\mu,\nu)$. The function $T^{\mathrm{a}}(\mu,\nu)$ is known as a coherent transfer function (CTF) and is distinguished from the OTF for an incoherent system. It is also well known that the CTF agrees directly with the pupil function of the lens. This means that the design of a pupil function determines the properties of the CTF directly, and this physical nature provides the conventional spatial filtering in optical information processing. Use of this technique in SFV will be described briefly in Sect. 3.5.

Figure 3.3 shows the OTF and the CTF of an aberration-free circular lens. Note that a cutoff of the curve for an incoherent system occurs at frequency $\mu_{\mathrm{max}} = 2\mu_0$, whereas that for a coherent system occurs at μ_0. However, this does not mean that the incoherent system has twice the resolution of the coherent system [55]. The OTF decreases monotonically with increasing spatial frequency. This means a decrease in the image contrast. The CTF is unity in the entire passband up to μ_0 and, then at μ_0, suddenly drops to zero. The CTF of unity means no amplitude distortion in the image. When the pupil diameter is denoted by D, the cutoff frequency μ_{max} is given by

$$\mu_{\mathrm{max}} = \frac{D}{\lambda f} = \frac{1}{\lambda F} \; . \tag{3.20}$$

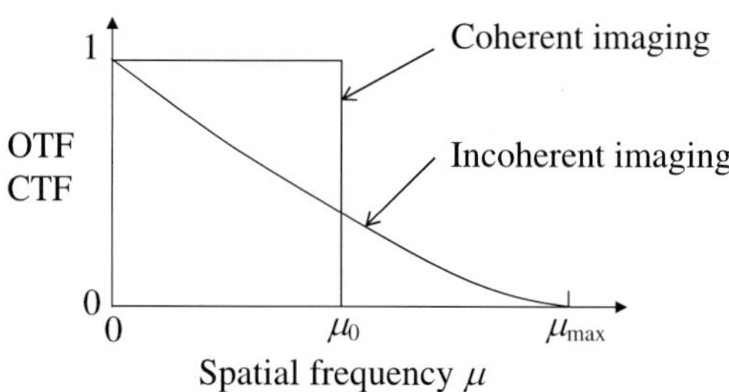

Fig. 3.3. Optical transfer function (OTF) and coherent transfer function (CTF) of an aberration-free circular lens

Thus, a larger diameter of the lens pupil transfers higher frequency components and gives higher resolution. As illustrated in Fig. 3.4, the movement of a low-contrast image on the grating produces a small intensity modulation relative to the dc component and, then, the output signal is of low visibility. It is found from Fig. 3.3 that this effect is motivated for higher spatial frequency components in the incoherent system. The resolution should be carefully considered for incoherent imaging, for example, of small particles and/or with small magnification. This is basically the same format as that discussed for point spread based on the Airy disk in Sect. 3.1.1. The spatial frequency that corresponds to the Rayleigh criterion or the radius of the Airy disk is written, by using (3.1), as

$$\mu_a = \frac{1}{r_a} = \frac{1}{1.22\lambda F} = \frac{1}{1.22}\mu_{\max} , \qquad (3.21)$$

which is called resolving power. It is found that the resolving power based on the Airy disk gives a cutoff frequency lower than the cutoff $\mu_{\max}$ of the OTF by 18%. This means that the Rayleigh limit is a slightly more rigorous criterion for the resolution than the cutoff of the OTF.

Let us consider an example of the F-number for the necessary resolution. The discussion of the size of small particles in Sect. 2.7.1 shows that the minimum ratio of particle image diameter to grating line interval, $2b/p$, is about 0.05. Then, the optical system should produce a particle image with a diameter $2b = 0.05p$ at its smallest with acceptable contrast. The OTF for an incoherent imaging system with an aberration-free lens having a circular pupil function is zero at the cutoff frequency $\mu_{\max} = 1/\lambda F$, as shown in Fig. 3.3. To ensure a certain degree of contrast, we should take $\mu_c = \mu_{\max}/2$, for example, as the upper limit of the available frequency range. Analogously to the Rayleigh criterion for resolution given by the radius of the Airy disk, we assume that radius b of a minimum particle image is the resolution limit. Then, a condition for obtaining the particle image may be written as

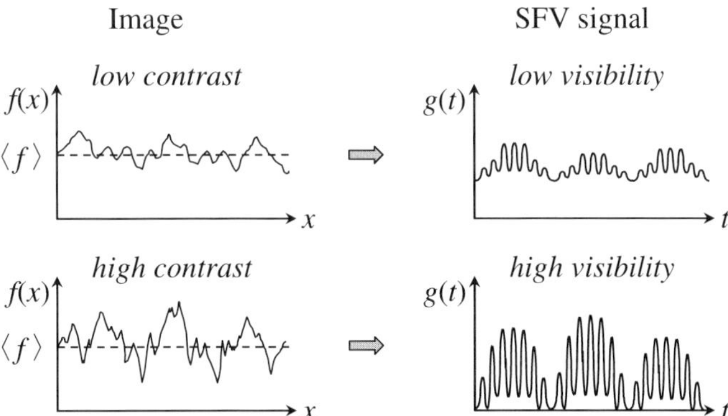

Fig. 3.4. Relation between the image contrast and the intensity modulation or visibility of output signals

Table 3.3. Calculated values of the acceptable cutoff μ_c and the corresponding F-number for a given line interval p of a grating with wavelength $\lambda = 500\,\text{nm}$

$p(\mu\text{m})$	$b(\mu\text{m})$	μ_c (lines/mm)	F
50	1.25	800	1.25
100	2.5	400	2.5
200	5	200	5
500	12.5	80	12.5
1000	25	40	25

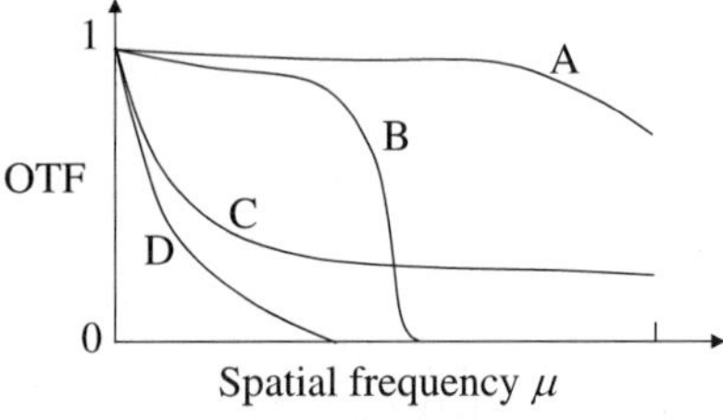

Fig. 3.5. Typical curves of the optical transfer function (OTF)

$$\mu_c = \frac{\mu_{\max}}{2} = \frac{1}{2\lambda F} = \frac{1}{b} = \frac{2}{0.05p} . \tag{3.22}$$

Table 3.3 shows numerical examples of μ_c and F for given values of grating line interval p with $\lambda = 500\,\text{nm}$. For the use of a grating with $p = 100\,\mu\text{m}$, the formation of the minimum particle image with $b = 2.5\,\mu\text{m}$ requires a resolution of $\mu_c = 400\,\text{lines/mm}$, which can be obtained from an F-number of 2.5.

The OTF is useful for understanding qualitatively the typical characteristics of an imaging system. Figure 3.5 illustrates four different curves for the OTF: curve **A** for high resolution and high contrast, curve **B** for low resolution and high contrast, curve **C** for high resolution and low contrast, and curve **D** for low resolution and low contrast. The OTF is also used for estimating the effects of lens aberrations and defocusing in optical systems, which will be described later in this chapter. Thus, the OTF is a better means than the Rayleigh criterion if the optical system is required to have a high degree of imaging performance in SFV measurements.

3.2 Lens Aberrations

In the above discussion on diffraction limited imaging, we assumed that the optical system is free from lens aberrations. Now we discuss briefly the effects of lens aberrations on SFV measurements. Aberrations mean departures of the exit-pupil wavefront from the ideal spherical form [55], and it generally becomes noticeable as the pupil diameter increases. When wavefront errors exist, light rays coming from a point in the object plane do not converge to an ideal geometric point in the image plane. The resultant image may be blurred or distorted, loses its sharpness and contrast and, thus, the signal visibility and the SNR are reduced. Lens aberrations are divided into two types: primary and chromatic aberrations. The primary aberrations are also called Seidel's five aberrations. The effects of these aberrations usually occur together.

3.2.1 Primary Aberrations

Primary aberrations include five different types: spherical, coma, astigmatic, field curvature, and distortion. For a complete description of these aberrations, readers may refer to textbooks on optics [66, 68, 69]. In Fig. 3.6, r_p and y indicate the radius of the exit pupil and the height of a point image on the y axis of an ideal geometric image plane. The mathematical analysis relates the five aberrations to r_p and/or y as shown in Table 3.4, together with their effects on image quality.

The spherical and coma aberrations are strongly influenced by pupil radius r_p whereas the other three are affected by the image height y. For a large pupil or small F-number, the effects of spherical and coma aberrations should be carefully be considered on image formation. A combination of convex and concave lenses is effective for reducing spherical aberration, and coma is improved by following Abbe's sine condition [66]. When the imaging system has astigmatic aberration, meridional and sagittal rays are focused onto different points along the optical z axis. There is usually the smallest cross-sectional circle of the imaging light flux, which is called "a circle of least confusion,"

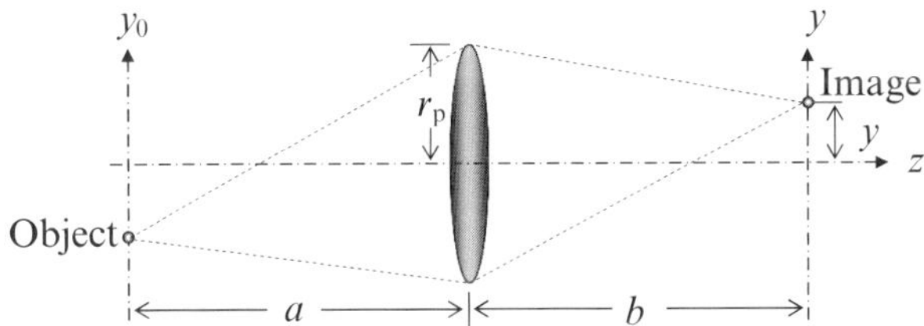

Fig. 3.6. Single lens imaging system for consideration of primary aberrations

Table 3.4. Effects of the five primary aberrations

Five aberrations	Dependence on r_p and y	Effects on image quality
Spherical	$\propto r_\mathrm{p}^3$	Reduction of contrast and resolution
Coma	$\propto r_\mathrm{p}^2 y$	Reduction of contrast and resolution
Astigmatic	$\propto r_\mathrm{p} y^2$	Image deterioration
Field curvature	$\propto r_\mathrm{p} y^2$	Reduction of contrast and resolution
Distortion	$\propto y^3$	Nonuniform magnification

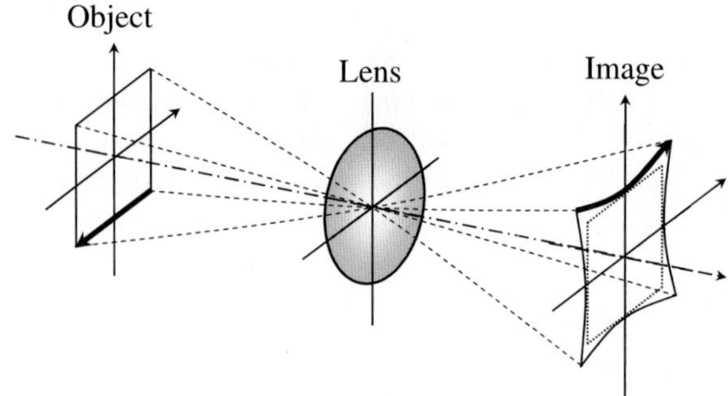

Fig. 3.7. Effect of distortion in an optical imaging system

between the two focusing points. The spatial filter may be placed at position z where this circle appears, as a point of compromise if astigmatic aberration is influential. Field curvature defeats plane-to-plane imaging, and a set of focused points result in a curved image. This curved surface causes focusing errors on the geometric image plane. The above four aberrations may result in a decrease in sharpness and contrast of images, unless their effects are negligible.

With an aberration of distortion, the image is distorted in proportion to y^3 in the geometric image plane, though sharpness is maintained. By this effect, the lateral magnification of imaging is more distorted in the area with larger y (far from the optical axis) in the image plane. From this defect, a linear movement of the point object with constant velocity may result in a distorted locus of the point image as shown in Fig. 3.7. This means that the image velocity is not observed constantly in the entire image area and causes errors in the measured velocity. Therefore, distortion may have a direct effect on the accuracy of SFV measurements, and it should be reduced preferentially among the five aberrations. The quantity of distortion can be evaluated by the expression

$$\text{Dist} = \frac{y' - y}{y} \times 100\% \,, \tag{3.23}$$

where y and y' denote heights of the point of interest in the ideal geometric and the corresponding distorted images, respectively. From the empirical point of view, a distortion of $\leq \pm$ 1–2% may be desired and that of $< \pm 5\,\%$ may be acceptable as a moderate criterion.

3.2.2 Chromatic Aberrations

The refractive index of lens materials (usually glasses) depends on the wavelength of light, and the focal length of lenses does also. The wave-front error due to this wavelength dependence is called a chromatic aberration. For SFV measurements using natural light or the illumination of white light, the chromatic aberration should be taken into consideration. As shown in Fig. 3.8a,

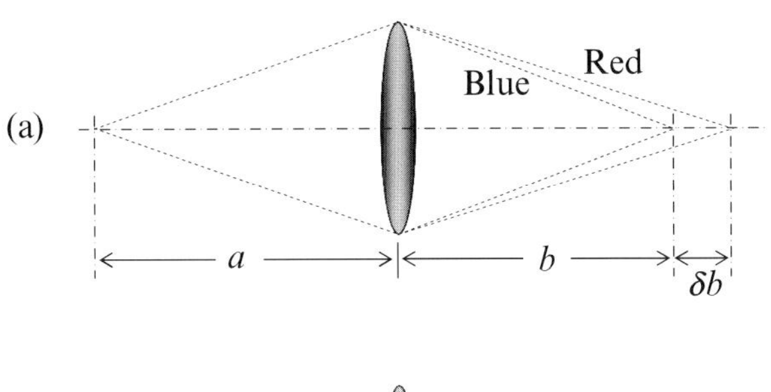

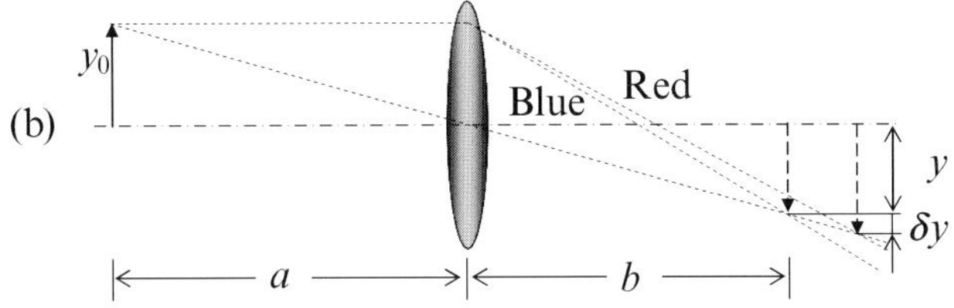

Fig. 3.8. Effects of (a) longitudinal and (b) lateral chromatic aberrations

the imaging on the optical axis is accompanied by a longitudinal chromatic aberration in which focusing points are separated on the optical axis for different wavelength components of light. This aberration is evaluated by the separation "δb" between two focusing points for the red and blue components. Imaging off the optical axis involves not only a longitudinal chromatic aberration but also a lateral chromatic aberration as shown by "δy" in Fig. 3.8b. The longitudinal error δb can be the cause of a blurred image and reduces the contrast, whereas the lateral error δy means a deviation of the lateral magnification from the ideal value and leads directly to the dispersion of measured velocity data.

The change δf in the focal length f of the lens due to a wavelength change is expressed by

$$\delta f = -\frac{f}{\nu_d} , \qquad (3.24)$$

where ν_d is the quantity known as Abbe's number and is given, for the visible wavelength range, by

$$\nu_d = \frac{n_d - 1}{n_F - n_c} . \qquad (3.25)$$

In the above equation, n_F, n_d, and n_c denote the refractive indexes at wavelengths of 486.1, 587.6, and 656.3 nm, respectively. Abbe's number ν_d can generally be referenced as one of the useful constants for characterizing optical materials. By using δf in (3.24), the longitudinal and lateral chromatic aberrations δb and δy are roughly estimated by simple geometric treatments as follows:

$$\delta b = \left(\frac{b}{f}\right)^2 \delta f , \qquad (3.26)$$

Table 3.5. Effects of the chromatic aberrations

Case	a(mm)	b(mm)	$M = b/a$	y_0(mm)	δb(mm)	δy(mm)	δM
(a)	200	200	1	10	6.23	0.01	0.001
(b)	110	1100	10	0.1	189	0.171	1.71
(c)	1100	110	0.1	10	1.88	0.0171	0.00171

$$\delta y = y_0 \delta M = y_0 \frac{bM}{f^2} \delta f \ , \tag{3.27}$$

where b is the distance between the lens and the ideal image plane without aberration, y_0 is the object height from the optical axis, $M = b/a$ is the lateral magnification of imaging, and δM is the deviation of lateral magnification due to δf. For example, a lens of BK7 with $n_{\mathrm{d}} = 1.5168$ and $\nu_{\mathrm{d}} = 64.2$ involves $\delta f \simeq 1.56\,\mathrm{mm}$ for a focal length of $100\,\mathrm{mm}$. With this condition, typical values for δb, δy, and δM are calculated in three simple cases, as given in Table 3.5. It is found from **(b)** that a large magnification may cause significant influence of chromatic aberrations. An appropriate color filter can be introduced to reduce the effects of chromatic aberrations, though the light intensity detected is decreased. Use of an achromatic lens usually gives better results.

According to empirical knowledge, SFV measurements are not significantly influenced by primary and chromatic aberrations, apart from artificial defocusing errors. Some special cases such as microscopic, telescopic, and wide-angle imaging need countermeasure, to lens aberrations. A lens design using ray tracing is required for an optical system for spatial filtering velocimetry.

3.3 Focusing Depth and Probe Volume

Even if the effects of diffraction and lens aberrations are negligible, an artificial focusing error is always a practical problem to be considered in SFV measurements. This section treats defocusing effects in an ideal imaging system under paraxial approximation, in which the depth of focus defines a probe volume depth in the SFV imaging system.

3.3.1 Depth of Focus

In a paraxial approximation, point-to-point imaging is realized in the geometric focal plane. With a focusing error, however, the image of a point object is spread to a circular spot. In SFV measurements, this problem appears when the moving object has a velocity component in the direction along the optical axis. In this case, the focal plane is displaced from the plane of the spatial filter, and the imaging equation of (3.3) is not satisfied. The defocused image is of low contrast and becomes a source of low SNR in output signals in the same way as lens aberrations.

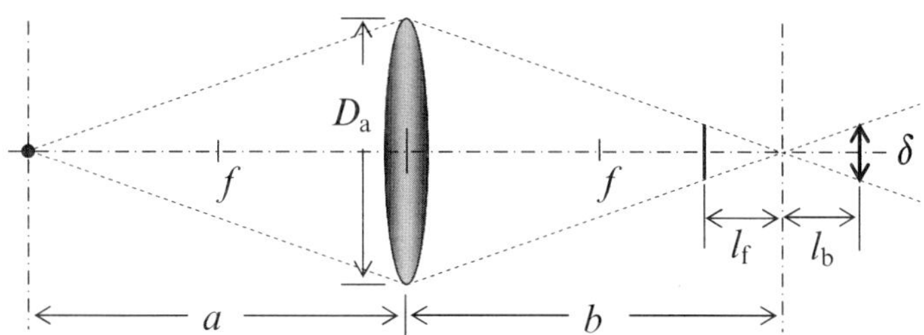

Fig. 3.9. Depth of focus in the optical imaging system

Figure 3.9 shows a simple imaging system using a single lens having a diameter D_a and a focal length f. Let us assume that δ is the maximum permissible diameter of the point spread due to an out-of-focus. The simple geometry in the figure gives a certain longitudinal range along the optical axis in which the defocused point spread is smaller than the maximum permissible diameter δ. This range is acceptable for focusing and is referred to as the depth of focus in an SFV optical imaging system. In Fig. 3.9, l_f and l_b denote the forward and backward depths of focus, respectively, and are given by

$$l_f = l_b = \delta \frac{b}{D_a} = \delta F_{eff} \ . \tag{3.28}$$

Reducing the lens aperture D_a increases the depth of focus. The focusing depth obtained in a real imaging system usually becomes smaller than the value estimated in (3.28) because of the lens aberrations involved. A large focusing depth contributes to the capacity of the optical system for axial movement of an object. A value of δ may be evaluated by considering the size effects discussed in Sect. 2.6, and is given roughly, for the grating line interval p by

$$\delta \simeq \frac{p}{2} \ . \tag{3.29}$$

This condition may guarantee both the acceptable visibility and the negligible deviation of the central frequency in output signals. With a lens of $D_a = 25\,\mathrm{mm}$, $f = 100\,\mathrm{mm}$, and $p = 1\,\mathrm{mm}$, the depth of focus $l = l_f = l_b$ is calculated as $4\,\mathrm{mm}$ for an imaging magnification $M = 1$.

In relation to the above-mentioned focusing, the design specifications of a telecentric imaging system have been theoretically and experimentally investigated [70] for SFV measurements. The results give useful information on the decision as to whether a single or a double telecentric system shall be used, the choice of the optical magnification, and the dimensioning of the aperture stop.

3.3.2 Probe Volume

In an SFV optical configuration, the probe volume is defined by both the illuminating light flux and the imaging system. Because the illumination will be described in the next section, this subsection treats the probe volume specified by the imaging system in which the object space is assumed to be uniformly illuminated.

On one hand, the cross-sectional area of the probe volume is defined by the optical projection of the window of the spatial filter on the object plane. For example, a circular window with a diameter of $2a$ in the image plane gives a circular probe cross-sectional area with a diameter

$$d_{\mathrm{v}} = \frac{2a}{M} \; .$$

(3.30)

A small probe area is advantageous for high spatial resolution, but with it the light being detected becomes weak and the number of grating lines, n, in the spatial filter can be limited to a small number.

On the other hand, the depth of the probe volume is defined by the focusing capacity of the imaging system. In Fig. 3.10, the following imaging equation holds with given notations,

$$\frac{1}{a} + \frac{1}{b} = \frac{1}{f} \; ,$$

(3.31)

and the ratio $M = b/a$ gives the lateral magnification. Any axial displacement of the point object from the initial object plane (distance a from the lens) produces a point spread due to defocusing in the initial image plane (distance b from the lens). Thus, there is another longitudinal range along the optical axis in the object space, in which the point is imaged as a point spread smaller than the circular spot having a maximum permissible diameter δ in the initial image plane. This range is an alternate to the focusing depth and is referred to as the depth of field. The forward and backward depths of field, t_{f} and t_{b}, in Fig. 3.10 are given geometrically by

$$t_{\mathrm{f}} = \frac{\delta F_{\mathrm{eff}} a^2}{f^2 + \delta F_{\mathrm{eff}} a} \; ,$$

(3.32)

$$t_{\mathrm{b}} = \frac{\delta F_{\mathrm{eff}} a^2}{f^2 - \delta F_{\mathrm{eff}} a} \; .$$

(3.33)

Usually, the following approximation is practically convenient with an acceptable error:

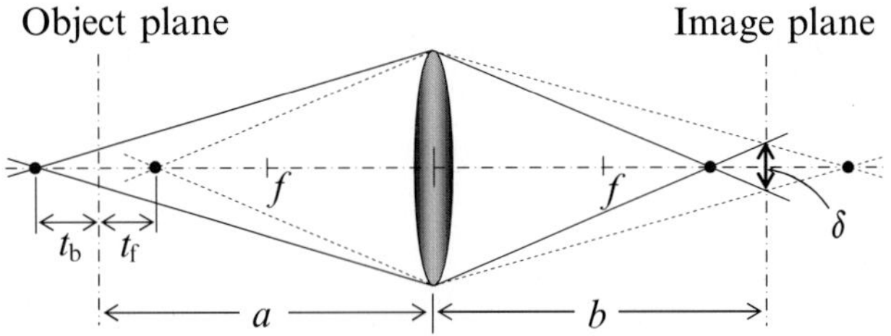

Fig. 3.10. Depth of field in an optical imaging system

$$t_{\mathrm{f}} \approx t_{\mathrm{b}} \approx t = \frac{\delta F_{\mathrm{eff}}}{M^2} , \tag{3.34}$$

where M^2 expresses the longitudinal magnification. Since the value of δ given by (3.29) may generate output signals with acceptable visibility in SFV, the range of $\pm t$ around the initial object plane is considered effective as the object space of the SFV imaging system. Thus, the probe volume depth V_{D} can be given by

$$V_{\mathrm{D}} = \pm \frac{\delta F_{\mathrm{eff}}}{M^2} . \tag{3.35}$$

The probe volume depth should carefully be considered in SFV measurements of fluid flows in which scattering particles are distributed three-dimensionally and also have three-dimensional velocity components. The large depth of field makes the SFV system insensitive to the longitudinal movement of particles. Conversely, axial resolution of the measuring position is obtained by a small depth of field. Three-dimensional objects, surfaces with large roughness or steps, and objects with mechanical vibration or fluctuation along the optical axis may also be probed by enlarging the probe volume depth sufficiently.

3.4 Illumination

Since the SFV technique processes the optical image of a moving object, measurements may be done in principle under ambient or natural light. An effective method of illumination, however, provides improvements in image contrast and the definition of the probe volume. Measurements in a microscopic region also require active illumination. Here we consider the illumination of SFV for two types of objects, particles in a fluid and rough surfaces, with some additional descriptions of coherent and incoherent light.

3.4.1 Small Particles in a Fluid

For measurements of small particles in a fluid, it is desirable to detect only the light scattered by particles and to suppress the background light. To realize this, dark-field illumination is usually employed. Figure 3.11 schematically shows the relation of the illuminating light flux and the optical imaging system. The collimated light flux having a width w_{l} illuminates the object with an angle θ to the optical axis of the imaging system. For a smaller angle of θ, the intensity of light scattered by particles is relatively large in the direction of the imaging system because this direction lies in the near-forward scattering scheme with respect to the illuminating light. However, a fraction of the illuminating light might directly come into the pupil of an imaging lens and result in a source of dc noise. The minimum value of angle θ that is able to eliminate this problem gives a lower limit of θ for dark-field illumination. From the empirical point of view, angle θ is usually set at 30–60°. The width w_{l} of the

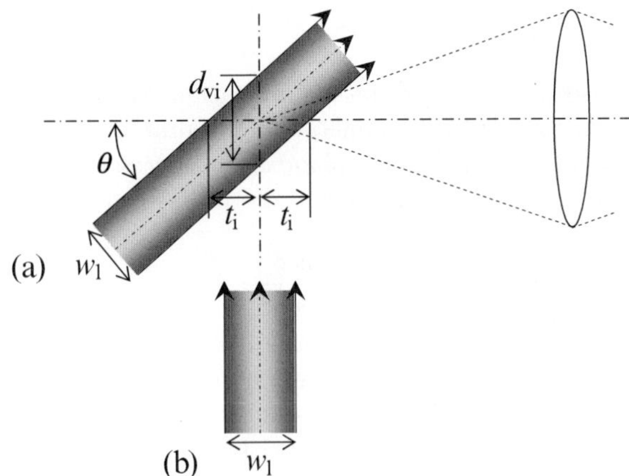

Fig. 3.11. (a) Illuminating light flux having a width of w_l with an angle θ to the axis of the optical imaging system, and (b) the side illumination scheme

illuminating light flux is closely related to the definition of the probe volume. From the geometry in Fig. 3.11a, the cross-sectional diameter d_{vi} and depth $2t_i$ of the illuminated volume with respect to the direction of the imaging system are given by

$$d_{vi} = \frac{w_l}{\cos \theta} \,, \tag{3.36}$$

$$2t_i = \frac{w_l}{\sin \theta} \,. \tag{3.37}$$

There can be two approaches for the definition of probe volume: adjustments of the illuminating and imaging systems. Usually, an adjustment of the optical imaging system may be a primary means to define the probe volume, because it is relatively easier and more deterministic than that of the illuminating system. In this case, the illuminated volume defined by (3.36) and (3.37) should be larger than the probe volume defined by (3.30) and (3.34). Thus, $d_{vi} > d_v$ and $2t_i > 2t$ are required. In some cases, the illuminating light may be required to define primarily the probe volume. Then, $d_{vi} < d_v$ and $2t_i < 2t$ should be satisfied. If the probe volume depth and cross section must be defined more precisely, side illumination or $\theta = 90°$, as shown in the illumination geometry (b) of Fig. 3.11, is preferable. In this illumination scheme, the depth of the probe volume is defined directly by width w_l ($< 2t$) of the illuminating light flux, whereas the probe cross-sectional area is defined by the imaging magnification M and the window diameter $2a$, or (3.30). In particular, the use of a light sheet in the side illumination, which is the standard illuminating means in particle image velocimetry (PIV) [3], is quite effective for high spatial resolution in depth.

In any case, it is desired that the intensity distribution in the illuminating light flux be as uniform as possible. For this purpose, Köhler illumination [66], which is often employed for microscopes, may be useful when white light is used.

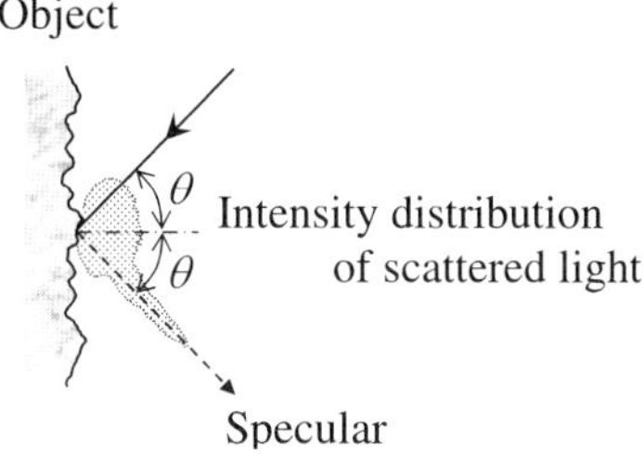

Fig. 3.12. Scattering of light at a rough surface

3.4.2 Rough Surfaces

For measurements of light-diffusing or reflective objects and rough surfaces in motion, the light reflected by the surface under illumination is usually employed for imaging. In Fig. 3.12, the surface is illuminated with an incident angle θ to the normal to the surface. The scattered light is diffused in all the directions, but the specular component having a large intensity is reflected toward the opposite direction at the same angle as incident angle θ. This strong reflection usually reduces the contrast of the surface image. Thus, the imaging optical axis should be taken out of the direction of the specular component. Specular reflection arises in general cases since the surfaces of reflective objects are not usually perfect diffusers. For rough surfaces, the probe cross-sectional area is simply defined by the imaging magnification and the window of the spatial filter.

3.4.3 Coherent and Incoherent Illumination

The SFV technique can employ both coherent and incoherent light sources for illuminating an object being measured. This freedom is a practical advantage in comparison with LDV. A choice of one source from the two light sources may be made by considering various aspects of the measurements.

On one hand, the use of the coherent light or laser light generally has the following merits for SFV measurements:

- Light collectivity with the high intensity is useful for measurements with high spatial resolution such as microscopic probing.
- High directivity is effective for specifying the measuring point required in remote sensing applications.
- Monochromaticity is advantageous for discriminating signal light from ambient or background white light and for improving image quality. It is also favorable for designing lenses and optical systems without consideration of the chromatic aberration.
- Polarization is effective for suppressing background light.

- The compactness of a laser diode is favorable for the configuration of a practical system.
- Coherent optical processing is available for modifying or improving image quality.

Fluctuations in the wavelength and intensity of a laser diode due to temperature change are not considered to cause serious errors to SFV measurements. On the other hand, the disadvantages of using laser light are as follows:

- High coherency causes a speckling phenomenon in the image of objects and may reduce the image quality.
- Ringing may appear at sharp edges or knife edges in the image of objects [55], resulting in an ambiguous border of the object in the image. The ringing means an attenuating oscillation in the intensity distribution near the edges and, then, a detection of the moving image containing the ringing may produce an unwelcome higher harmonic noise.
- Highly coherent and strong illumination is quite sensitive to optical imperfections [55]. For example, tiny dust particles in an optical system may lead to undesirable diffraction patterns and produce scattered dc light which is superimposed on the image.

An incoherent source can be simply and widely used for general cases of SFV measurements. It may be better than a laser source when a relatively wide area should be illuminated. For SFV measurements with a microscope, incoherent illumination is usually employed. Use of an incoherent source, however, means that the effects of chromatic aberrations should be considered for some cases which require high accuracy in measurements.

As discussed with Fig. 3.3 in Sect. 3.1.2, a comparison of the OTF and CTF does not simply mean that incoherent illumination will give higher quality in an image than coherent illumination. This comparison is generally far more complex than one aspect of examinations suggested by the graph of Fig. 3.3. Imaging with coherent illumination is largely influenced by the phase distribution of the light scattered by the object, and the problem of whether the image quality is better or worse than that with incoherent illumination depends on the case. It is quite difficult to state which type of illumination is to be desired in all cases. Therefore, it may be recommended that, for a choice of one source from the two types of illumination, we should take account of the above-mentioned features and estimate whether or not they satisfy the requirements and circumstances for measurements.

3.5 Image Modification

3.5.1 Spatial Frequency Filtering

In coherent optical information processing, there is a well-known technique for modifying the image, called spatial frequency filtering. This technique

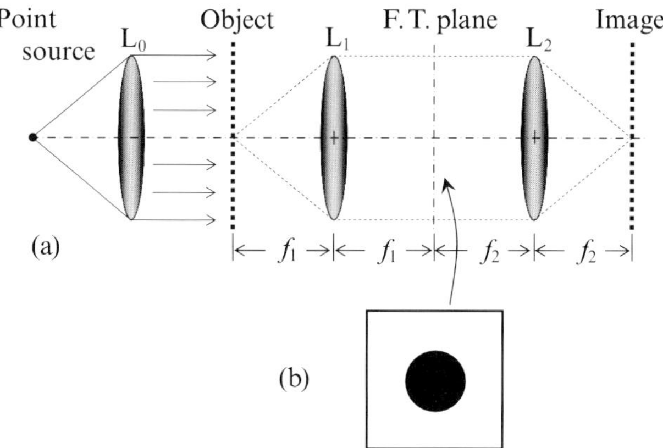

Fig. 3.13. (a) Two-lens imaging system for optical spatial frequency filtering, and (b) a typical high-pass filter

is named by the same term "spatial filtering" as that of spatial filtering velocimetry, the main subject of this book. It should be, thus, noted that spatial filtering for coherent image processing described in this subsection is different from that for velocity measurements giving the principle of the SFV, although the optical effect or the filtering operation of an input image in the spatial frequency domain is common between the above two cases of spatial filtering.

Figure 3.13a illustrates a two-lens imaging system for spatial filtering, where an object is illuminated by a plane wave generated from a point source on the optical axis. There is a Fourier-transform (FT) plane at the focal distance f_1 forward from lens L_1 and at the focal distance f_2 backward from lens L_2. In the FT plane, we can observe an intensity pattern which is proportional to the Fourier transform of the object intensity distribution. Lower spatial frequency components of propagated light pass through the FT plane near the optical axis or the center, whereas higher spatial frequency components pass through the area far from the optical axis in the FT plane. Then, the use of a mask in the FT plane realizes spatial frequency processing of the input intensity pattern. For example, a simple opaque spot centered in the FT plane shown in Fig. 3.13b functions as a high-pass filter, which suppresses dc and low-frequency components of the object intensity pattern and, then, is often effective for improving image contrast and for enhancing the edges of the image. This image modification may improve the signal visibility. There are some other spatial filters available for reducing noise, sharpening, and so on [71]. Note that monochromatic plane-wave illumination is important in using the spatial filtering technique. Illumination by a collimated or focused laser beam in the axial direction nearly satisfies this condition. Therefore, when this type of illumination is used, the spatial filtering technique is potentially useful for modifying the image in SFV measurements.

3.5.2 Photographic Filters

In photography, various filter products are commercially available. They can be generally used for controlling and compensating for color characteristics. But they are also useful for removing surface-reflected light, for enhancing image contrast, and for creating various artificial effects in the image. Although these filters are basically designed on the assumption that images are recorded on photographic films, some of them can also be useful for modifying the image in an SFV optical imaging system. For example, a polarizing filter is sometimes useful for suppressing background light and undesirable light reflected from surfaces. The polarizing filter transmits only the light component in the polarizing direction of the filter and blocks its orthogonal component. Thus, an appropriate orientation of the polarizing direction in the filter may propagate light coming from the object and suppress the other unwelcome light. There are other filters available for improving the image contrast in SFV measurements.

4

Signal Analysis

In the SFV technique, velocity information is derived from frequency measurements of output signals from a photodetector. To transform the raw signals into velocity with better accuracy, appropriate electronic equipment is required. As described in Chap. 2, the photodetector output in SFV is a narrow-band random signal whose central frequency is proportional to the object's velocity. In principle, the central frequency may be measured by simple frequency counters or spectrum analyzers. But this equipment is often unsatisfactory for real SFV signals, which contain random amplitude and phase fluctuations, sometimes with a poor signal-to-noise ratio. Fortunately, this type of signal is quite similar to that in the LDV technique, and various equipment that has been developed for processing LDV signals is available for SFV signals. This means that no special signal-analyzing instrument must be newly developed at least for usual SFV measurements and, in this respect, the SFV technique is favorably situated. Typical equipments for LDV measurements and their related studies are well reviewed in the literature of Durst et al. [1], Durrani and Greated [2], and Drain [5].

This chapter is devoted to the consideration of signal-analyzing techniques used for SFV measurements. Types of output signal peculiar to SFV measurements are first discussed in Sect. 4.1. In Sects. 4.2–4.5, typical signal-analyzing techniques are briefly introduced, including spectrum analysis, frequency tracking, counting techniques, and correlation analysis. Section 4.6 summarizes some guidelines for choosing an appropriate technique for given conditions.

4.1 Types of SFV Signals

To choose an appropriate signal-analyzing technique, consideration must first be given to the types of SFV signals. Continuous periodic signals with a constant amplitude are easily processed by popular equipment such as simple frequency counters or spectrum analyzers, if the signal-to-noise ratio is not

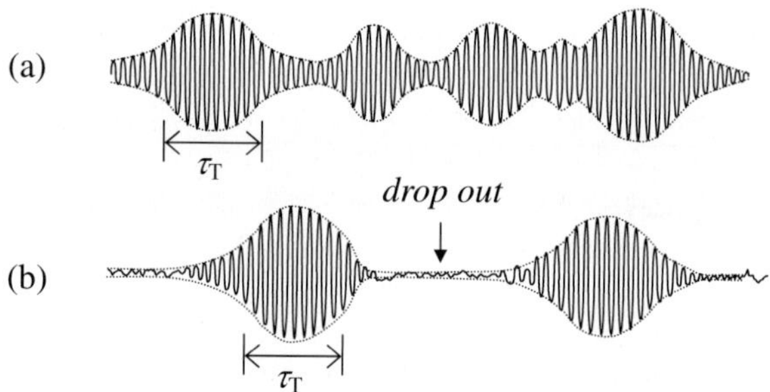

Fig. 4.1. Two types of the pedestal-eliminated SFV signal; (**a**) continuous with a random amplitude and (**b**) intermittent

poor. No significant discussion is needed in this case. However, this type of signal is very rare in SFV measurements. The situation that produces such a rather ideal signal must be that images with the same total intensities continue indefinitely to pass through the spatial filter without interruption. Significant consideration may be necessary for two other types of pedestal-eliminated signals depicted in Fig. 4.1: (**a**) continuous periodic signals with a random amplitude and (**b**) intermittent signals or burst-like periodic signals.

The signal of type (**a**) in Fig. 4.1 occurs when the surface pattern of a large or long object, a speckle pattern, or many particles in a fluid are measured. The contribution from each bright spot in the surface image or the speckle pattern may be equivalently treated in the present discussion to that from a particle image. The randomness of the amplitude envelope is due to the addition of contributions from individual particle images which are distributed in a disordered manner over the spatial filter window in the image plane. Each particle image passing over the window area yields a burst of periodic signal lasting a time τ_T, the "transit time" in that area, as shown in Fig. 4.1b. The time τ_T can be estimated by $\approx 2a/v$, where $2a$ is the size of the spatial filter window and v is the velocity of particle images. The addition of this type of signal from many uncorrelated particle images produces random amplitude and phase fluctuations. After passing of a certain particle image, the next burst is generated from a completely different particle image. Therefore, the amplitude and phase fluctuations statistically contain the timescale of the transit time [1, 2, 4, 5]. This means that the corresponding frequency spectrum contains a broadening of $\Delta f \approx 1/\tau_T$. The spectral broadening causes ambiguity in frequency measurements and is often referred to as "ambiguity noise" [72]. In this case, problems can occur in the small amplitude between any two successive bursts of signals. The phase changes irregularly and the signal-to-noise ratio is momentarily degraded in this part of a signal. Special treatments may be necessary to deal with these problems.

When the particle concentration is low, output signals may be intermittent, as shown in Fig. 4.1b. Each burst of periodic signals is produced from the passage of each particle image through the window area of the spatial filter.

Such bursts occur intermittently with certain periods of no signal, which are sometimes referred to as "dropouts" [5, 73]. If equipment that is designed for continuous signals is used, the dropout periods are erroneously recorded as periods of signal oscillation, and the resultant frequency contains a large amount of error. Signal processing techniques are, thus, required to detect dropouts in their first stage and to operate only in the period when signals are present. Another problem should also be considered with the type of signal. If the interval of the signal bursts or the period of no signal is long compared with the timescale of velocity fluctuations in the object's movement, it is hardly possible to measure the changes in velocity. Limited signal processing instruments may have to be chosen for this case.

4.2 Spectral Analysis

One of the most popular techniques for frequency measurements is spectrum analysis. The major types of spectrum analyzers were formerly the frequency scanning and filter bank that were constructed mainly from analog circuits. Frequency scanning was widely used as a general-purpose instrument for processing LDV signals in the early development of LDV techniques. These days, a type using the fast Fourier transform (FFT) is also very widely used for spectrum analysis. Some books [1, 2, 5] on LDV techniques describe those types of instruments well.

4.2.1 Frequency Scanning

The frequency scanning type basically operates in the heterodyne mode. Figure 4.2 shows the block diagram of this type, which consists of a frequency mixer, frequency-variable oscillator controlled by a sweep generator, intermediate-frequency (IF) band-pass filter, and rectifier for squaring and smoothing operations. The input signal having a frequency f_0 is mixed with the oscillator output containing a frequency f_1 which is variable with an applied voltage from the sweep generator. The output from the mixer, which contains the frequency $f_1 - f_0$, is fed into the IF filter tuned to a fixed intermediate frequency f_{IF}. The filtered output is, then, rectified and smoothed. When the frequency f_1 of the oscillator is swept through a range of frequencies from f_{IF} to $f_{IF} + f_s$, the response of rectified output appears in the range of frequencies from 0 to f_s. This range is due to the frequency difference $f_1 - f_0$ which passes through the IF filter. By recording this response as a function of the variation of f_1 by an X–Y data set, the result presents a spectrum of photocurrent signals, showing a peak at a particular frequency $f_s = f_0$, which is proportional to the velocity of an object in SFV measurements.

This type of spectrum analyzer is still very popular as a standard instrument for spectrum analysis. A typical frequency range is from a few kHz to several GHz. The response bandwidth at a given frequency in the sweeping

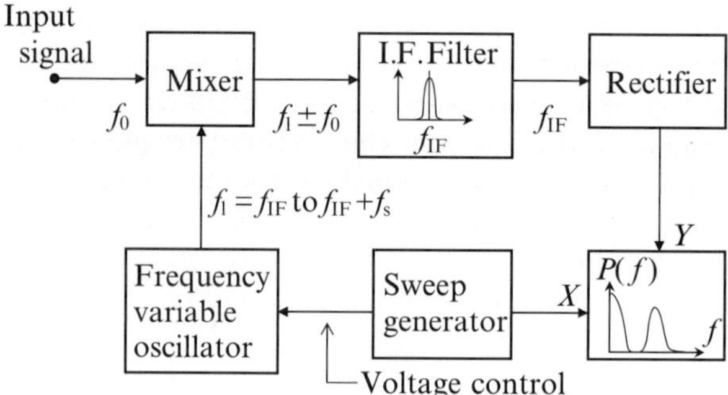

Fig. 4.2. Block diagram of a frequency-scanning type spectrum analyzer

band is determined by the passband properties of the IF filter. This is usually adjustable and about 10 Hz is a typical commercial value for the minimum of the response bandwidth. When the response bandwidth is given by Δf, it takes a time $1/\Delta f$ for the analyzer to respond. Thus, the problem with this instrument is that the time taken to sweep through the entire frequency range must be long. It is a fatal disadvantage for measurements of velocity with fluctuations. However, performance is being further improved by advanced electronics. The practical advantage is that this type can be widely used as a general-purpose instrument.

4.2.2 Filter Bank

Filter bank processing was developed to remove the necessity for frequency scanning. Instead of scanning the oscillator frequency for the use of a single band-pass filter, this type of instrument employs multichannel filters, i.e., a bank of simple band-pass filters which are tuned to different frequencies in the range. A spectrum of photocurrent signals can be built up from responses of the multiple filter circuits operating in parallel. It is, thus, obtained much more quickly than by frequency scanning. Generally, the frequency resolution of this instrument is relatively low because the number of filters is limited practically. The instrument also requires correct tuning of filters and uniform sensitivity for associated circuits. Integration of the responses of all the filters is effective for input signals with extremely poor SNR and intermittent signals. Since the tuning frequencies of individual filters are fixed, this type is unsuitable for general use.

4.2.3 Fast Fourier Transform

The photocurrent signal and its frequency spectrum are connected by a Fourier transform relation. The power spectrum, which corresponds to the result actually measured with a spectrum analyzer, is given by a squared absolute of the frequency spectrum. By executing the Fourier transform instrumentally, a different type of spectrum analyzer can be realized. For numerical computation of the Fourier transform for finite-time input signals, an extended version of the transform called "discrete Fourier transform" (DFT) is available. When a signal function and its Fourier transform are denoted by $g(t)$ and $S(f)$, respectively, the customary formulas giving their relation are (see Appendix A.2)

$$S(f) = \int_{-\infty}^{\infty} g(t) \exp\left(-\mathrm{i}2\pi f t\right) \mathrm{d}t \,, \tag{4.1}$$

$$g(t) = \int_{-\infty}^{\infty} S(f) \exp\left(\mathrm{i}2\pi f t\right) \mathrm{d}f \,. \tag{4.2}$$

The corresponding expressions in terms of DFT are

$$S(n) = \sum_{k=0}^{N-1} g(k) \exp\left(\frac{-\mathrm{i}2\pi n k}{N}\right) \,, \quad n = 0, 1, 2, \ldots, N-1 \,, \tag{4.3}$$

$$g(k) = \frac{1}{N} \sum_{n=0}^{N-1} S(n) \exp\left(\frac{\mathrm{i}2\pi n k}{N}\right) \,, \quad k = 0, 1, 2, \ldots, N-1 \,, \tag{4.4}$$

where n and k are a discrete frequency and discrete time for sampling the spectrum and the signal, respectively. N is the total number of sampling data for both signal and spectrum. If the sampling theorem is satisfied for the discrete spectrum and signal, the DFT can be a good approximation to the continuous Fourier transform.

Examination of (4.3) reveals that N^2 complex multiplications and $N(N-1)$ complex additions are required to implement the computation of the spectrum. This means a dramatic increase computation time with a larger number of samples N. In 1965, Cooley and Tukey presented a fast Fourier transform (FFT) algorithm, which computes the discrete Fourier transform much more rapidly than direct computation algorithms. For the details of the FFT algorithm, readers may consult the books by Brigham [74] and Bracewell [75]. A noticeable point of the FFT algorithm is that it can reduce the number of complex multiplications from N^2 to $2N \log_2 N$ by dividing the entire computation into several steps.

A spectrum analyzer using the FFT is commercially available, usually providing two channel simultaneous analysis of 1024 to 2048-point digitized input signals containing frequencies from a few hundred kHz to several MHz. Figure 4.3 shows a block diagram of a typical FFT spectrum analyzer (only a single channel is shown). Input signals are amplified, filtered, and fed into an analog-to-digital converter (ADC). Digitized signal data are stored in a memory and, then, are sent to a digital signal processor (DSP), which executes the FFT under the control of a central processing unit (CPU). The resultant spectrum is displayed and also can be fed into external units such

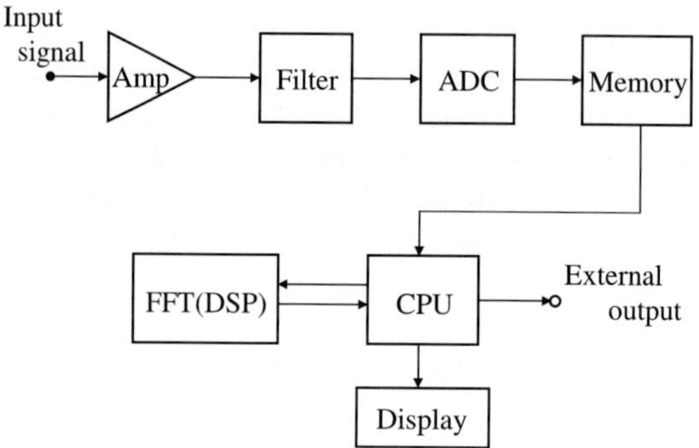

Fig. 4.3. Block diagram of a typical FFT spectrum analyzer

as an external memory, computer, printer, and so on. The typical resolution of current FFT processors is usually 16 or 32 bits. The real time operation is limited to input signals containing a frequency up to a few kHz due to the analog-to-digital conversion rate.

Due to the recent development of a high-speed microprocessor, FFT operation can also be executed in a personal computer (PC). There are many FFT source codes published in textbooks. In this case, the performance of the FFT analysis is determined by the sampling rate, the resolution of the ADC, and also the CPU rate and the memory size of the PC. There are also one-board type FFT signal processors that have been developed for LDV measurements. This type of processor can be used directly for treating SFV signals due to quite similar types of photocurrent signals. The processor board is inserted into a slot of the PC and, then, the PC is able to work as a convenient FFT-type signal processor without any additional instruments. The photocurrent signal is fed directly into the PC, which computes the spectrum and also offers postprocessing such as statistical analysis, graphic expression, database construction, and network transfer in its own framework. Thus, the one-board type offers compactness, convenience, and low cost with reasonable performance for SFV signal processing systems.

4.2.4 Maximum Entropy Method

The FFT is a basic and practically useful tool in almost every aspect of the spectral analysis of digital signals by computer. For accurate analysis, however, it requires at least the data record of several times of the signal period T_0, which is $1/f_0$ or the inverse of the central frequency being measured by the SFV principle described in Sect. 2.1. The Fourier transform of a short data record of signals less than a few period results in a large broadening of the spectrum, which means poor spectral resolution or an increase in errors in determining of the central frequency f_0. There is a nonlinear spectral analyzing method known as the maximum entropy method (MEM) [76]. The

MEM enables us to derive effective spectra from short data records and yields much better spectral resolution than the FFT. The method assumes no data outside the time interval specified [77].

In comparison with the FFT, the MEM generally enhances peak-like components of the spectrum, but sometimes spectrum splitting and peak frequency shifting happen for some types of signals. This is strongly dependent on the order of the prediction-error filter, which must be specified in the MEM computation. Some studies are useful for determination of the optimum order in the prediction-error filter. One useful criterion is

$$m < (2 \sim 3)\sqrt{N} ,\qquad(4.5)$$

where m is the order of the prediction-error filter and N is the number of sampled data [78, 79]. Improved MEMs have also been reported, including Fougere's nonlinear error minimization procedure [80]. The details of the MEM are not presented here since it is beyond the aim of this book. Some textbooks [77, 81] describe the basic theory, related properties, algorithm, and source codes for computer implementation of the MEM. Ingenious use of the MEM provides us with high-resolution analysis of the central frequency f_0 in SFV measurements. It may be particularly effective for the case of the small number n of grating lines in a spatial filter, with which the signal containing only a few periodic cycles is generated. Note that dispersion or broadening of the spectrum cannot be evaluated by the MEM, in principle.

4.3 Frequency Tracking

In measuring velocities with fluctuations, frequency tracking offers reliable signal processing. This type of processing follows the velocity fluctuation by monitoring the signal frequency itself. This means that the processing system automatically tracks the signal. To maintain tracking continuously, periodic signals must continue to appear. Thus, frequency tracking is good for moderate to comparatively high particle concentrations.

4.3.1 Frequency Tracker

Figure 4.4 shows the block diagram of a typical frequency tracker. The input signal containing a central frequency f_0 is mixed with output having a frequency f_1 of a voltage-controlled oscillator (VCO). The mixer output with a frequency $f_1 - f_0$ is applied to a band-pass filter tuned to a frequency f_c with a bandwidth of Δf_c. The output of this filter is fed into a frequency discriminator, which is a kind of frequency-to-voltage converter. If the frequency of the applied signal is exactly tuned to f_c, the discriminator circuit yields a zero output. Otherwise, the output contains a certain voltage e which is proportional to the frequency difference between f_c and $f_1 - f_0$. This output is then integrated and becomes the tracker output e_0, which is used to control

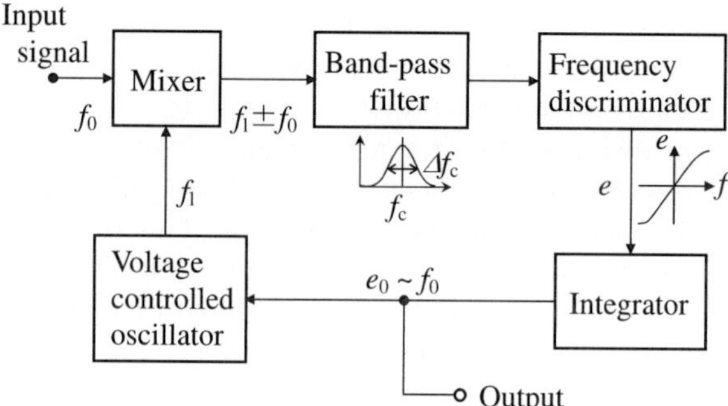

Fig. 4.4. Block diagram of a typical frequency tracker

the frequency f_1 of the VCO so that the frequency $f_1 - f_0$ agrees with f_c. In this way, the feedback loop is locked to the input signal. The output e_0 is in proportion to the input frequency f_0. Thus, the velocity fluctuation can be extracted from the frequency tracker as a continuous analog output.

Since this type of processing tracks the input signal, the feedback loop drifts out of control once the signal breaks off or drops out. To protect the circuit from dropout, usual trackers are designed to hold the preceding output during the dropout period and to lock in again as the signal comes in. However, this operation is often unsatisfactory for intermittent signals due to long dropout periods between signal bursts. Use of the frequency tracker may be problematic for low particle concentrations. The tracking performance depends on the bandwidth Δf_c of the band-pass filter and the time constant of the integrator. A good SNR is also desirable in the input signal for the frequency tracker. Otherwise, the circuit tracks noises and then drops out. This type of commercial equipment is available with various specifications to manage the dropout problem.

4.3.2 Autodyne

The autodyne system [82, 83] is another useful type of frequency tracking. In this system, a local oscillator is controlled so that its output frequency approaches that of the input signal. When the system is tracking, the frequency difference between the signal and the oscillator output is zero. This corresponds to the special case where the central frequency f_c of the band-pass filter is effectively zero in the conventional frequency tracker described in the previous subsection. To realize such a zero frequency difference, this system needs a discriminator to determine the sign of the frequency difference between the input signal and the oscillator output. Then, the voltage proportional to the frequency difference with its sign is fed back to control the oscillator to decrease the difference. Also, an additional unit is usually included to hold the oscillator output during signal dropouts.

The autodyne system has some advantages in comparison with the conventional frequency tracking system. The oscillator output can be used directly for measuring the average value of the SFV central frequency when the velocity fluctuates. Also because of the easily changeable integration time in the feedback loop, the system is rather flexible for different types of input signals. A narrow passband can be used to improve the output performance for the input signal with a poor SNR, whereas a wide bandwidth enhances the feedback response and is advantageous for quick recovery if the loop is broken.

4.4 Counting Techniques

For continuous periodic signals with good quality, the counting technique is the easiest way to determine the central frequency. If the signal burst can be appropriately detected, it is satisfactory even for intermittent signals. The pedestal or low frequency fluctuating component must be removed when this technique is applied. Counting is usually done for several cycles of the periodic amplitude oscillation in the signal. There is also an alternate technique to count the period of just one cycle. The counting technique is generally superior in processing time and accuracy to the spectrum analysis for determining the central frequency, but it is sensitive to noise. A good SNR is an important requirement for the input signal in applying the counting technique.

4.4.1 Frequency Counter

Originally, the frequency counter was equipment to count simply the number of cycles of a continuous periodic signal in a prescribed time. One cycle can be recognized by zero crossings of the signal amplitude. Thus, the equipment actually counts the number of zero crossings in the time, and the accuracy of one count is basically limited to $\pm$ a half cycle. The equipment may count zero crossings of noises unless a substantial signal exists. The above counting method is, therefore, inapplicable to intermittent signals when the time length of a signal burst is shorter than the prescribed time for counting.

For intermittent signals, it is better to count the number of high-frequency clocks in the period for the prescribed number of cycles within a single signal burst. This is schematically shown in Fig. 4.5. An order of GHz is a general example of clock frequency, which is very much higher than usual central frequencies of SFV signals. To detect the arrival of a signal burst and start the counting, it is usual to set a threshold at a certain nonzero level, as shown in Fig. 4.5. The threshold level must be determined so that it can adequately be above the noise level in the period of no signal. The counting is started as the signal amplitude exceeds the threshold and is stopped after the prescribed number of cycles is completed. The prescribed number should clearly be less

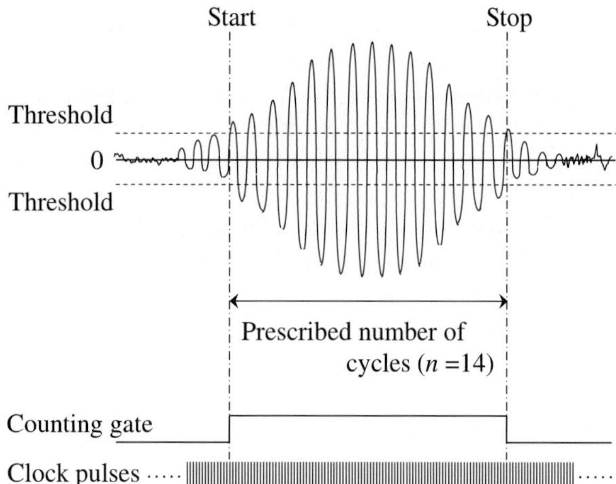

Fig. 4.5. Counting of clock pulses in a prescribed number of cycles within a single-burst signal

than the number of cycles contained in a single signal burst, which basically corresponds to the number of grating lines in the spatial filter. The total number of counted clock pulses gives a period for the prescribed number of cycles, and the average period obtained for one cycle is used to determine the central frequency. With this method, counting accuracy is determined by the clock frequency and, then, is substantially improved in comparison with the counting of cycles.

To guarantee the reliability of counting, commercial equipment usually employs some validation schemes [84] which reject the results of measurements if their accuracy is suspect. A typical scheme is to compare the two results of measurements obtained for different prescribed numbers of cycles. If the two results agree with each other within a predetermined error limit, they are accepted and, otherwise, rejected. For poor-quality signals, special attention should be paid to the validation scheme, including the threshold level and the prescribed number of cycles. A moderate application of validation may reduce the counting accuracy, whereas the strict validation decreases the efficiency of data acquisition. There may be a compromise between accuracy and efficiency. Unless the compromise condition can be found and settled for a signal of interest, it may be concluded that the quality of such a signal is insufficient for applying the counting technique.

4.4.2 Wave-Period Measurements

The number of cycles in a single signal burst can be decreased to less than 10 when the velocity in a small probing area is measured or when the measurement should be made with high spatial resolution. In this case, the counting techniques described above operate unsatisfactorily because the number of data rejected may increase due to insufficient numbers of cycles, and, then, it may take a long time to accumulate enough valid data to ensure the necessary accuracy. Those techniques are also problematic if the velocity fluctuation is contained during the single passage of a particle over the window area of the spatial filter. In these situations, the wave-period measuring technique [85] is an effective means to measure the instantaneous local velocity or the velocity fluctuation in a small probing area. The basic treatment is to obtain one frequency datum by measuring the period of one cycle. The signal burst having n cycles can produce, in principle, the n number of the frequency data. This processing is advantageous for increasing data acquisition efficiency and shortening measurement time. Because the instantaneous velocity is determined every cycle, the velocity fluctuation is, thus, measurable with this technique.

To realize the wave-period measuring technique effectively, high-speed processing circuits are required since, during the processing of data measured from a certain cycle, the equipment has to pause to measure the next cycle. The time for the processing results in dead time in measurements. Figure 4.6 shows a basic time chart of the wave-period measuring technique using the time-to-pulse height converter and the pulse height analyzer [86, 87]. The input photocurrent signal (**a**) is first amplified and fed into a band-pass filter to remove higher frequency noises, including shot noise and the pedestal component. The filtered signal (**b**) is then converted to the square wave form (**c**) by the Schmitt trigger circuit. The flip-flop circuit is next used to double the period of the square wave signal (**c**). During the period for a positive level of the resultant square wave signal (**d**), which corresponds to the period of one cycle of the input signal (**a**), a certain reference voltage is integrated. After the period, the integrated voltage (**e**) is maintained in the hold circuit to settle the reciprocal calculator, by which the inverse value of the integrated voltage

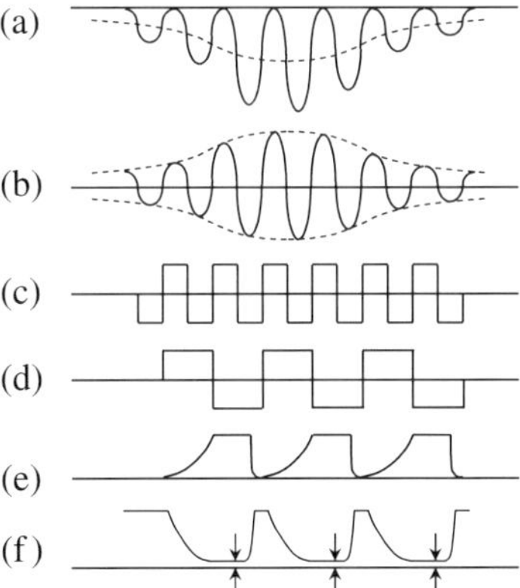

Fig. 4.6. Time chart of the wave-period measurement using the time-to-pulse height converter and the pulse height analyzer

is obtained. This output (**f**) is proportional to the central frequency of SFV signals. Finally, this is sampled and converted to digital data. In this example, the frequency measurement can be repeated every other period of the input signal. Alternate processing with two channels of the same circuit can realize measurements of all cycles in the burst.

Instead of using the time-to-pulse height analyzer, it is also possible to count high-frequency clock pulses for determining the period of one cycle. For this purpose, the clock frequency has to be much higher than the central frequency of SFV signals being measured. Thus, this clock-counting type may be ineffective for SFV signals containing a higher central frequency than a few MHz, whereas it is advantageous for digital processing techniques.

In the usual wave-period measuring method, a number of frequency data are fed into a statistical analyzer, from which a histogram of the frequency data is produced. The most probable frequency of the histogram can be regarded as the central frequency of signals and is finally used to calculate the objects velocity. The wave-period measuring technique is quite useful for intermittent signals and signals containing a small number of cycles (usually less than 10) in one burst, but it is less satisfactory for deteriorated signals with noise and/or phase fluctuations because measured results depend strongly on the zero-crossing properties of signals and the zero-crossing intervals can easily be affected by noise and wave distortions. This problem is theoretically investigated by the zero-crossing probability of signals [53,88,89]. The dispersion of measured frequency data decreases as the number n of grating lines (or the number of cycles in a single signal burst) and the SNR increase, and as the bandwidth of the band-pass filter becomes narrower [27].

A band-pass filter is a necessary device for the wave-period measuring technique and should be used properly. The passband of the filter should be narrower and its central frequency should be set as closely to the signal frequency as possible. A tracking-type band-pass filter is a useful means for following the variation of the signal frequency with a comparatively narrow passband [87]. The pedestal component can also be removed by an optical arrangement such as dual channel detection with a differential amplifier, which will be described in the next chapter. This optical means may reduce the load of the electric band-pass filter mentioned above.

4.5 Correlation Analysis

The correlation technique is another useful processing tool for determining the central frequency of SFV signals. Although the technique was originally effective for second-order statistical analysis of random variables, it is also available for the frequency determination of periodic signals. Mathematically, the temporal autocorrelation function $R(\tau)$ of photocurrent signal $g(t)$ is defined as

$$R\left(\tau\right) = \lim_{T \to \infty} \frac{1}{T} \int_{-T/2}^{T/2} g\left(t\right) g\left(t + \tau\right) \mathrm{d}t \;, \tag{4.6}$$

where T denotes a specified integration time which is sufficiently long compared to the period of a cycle of the signal. In general, the autocorrelation function $R(\tau)$ is related to the power spectrum $G_\mathrm{p}(f)$ of the signal $g(t)$ by a Fourier transform relation (Wiener–Khintchine theorem, see Appendix A.4). There are three typical methods for correlation analysis: the direct correlation technique of photocurrent signals, the FFT of the power spectrum, and the photon correlation technique. Generally, correlation analysis is quite effective for noisy signals, but it necessitates a long integration time to accumulate data sufficiently.

4.5.1 Autocorrelation of Photocurrent Signals

The basic processing in correlation analysis is to calculate the function $R(\tau)$ of (4.6) directly from signal $g(t)$. Let us consider a typical burst-signal model of a sinusoidal function with a Gaussian envelope, as shown in Fig. 4.7a, in which τ_T indicates the transit time of a particle image over the window of the spatial filter and T_0 is the signal period. The power spectrum $G_\mathrm{p}(f)$ of the signal (**a**) takes the form of Fig. 4.7b having a peak centered at the frequency $f_0(= 1/T_0)$. A typical form of the autocorrelation function for signal(**a**) is depicted in Fig. 4.7c. The oscillatory correlation function is gradually attenuated with the Gaussian envelope as the delay time τ increases. The period of this oscillation is T_0, or that of the signal. The "lifetime" (usually e^{-1} width) of the oscillation or the correlation time of the envelope function corresponds to τ_T, or the transit time. Thus, the central frequency f_0 may be determined by measuring the period T_0 of the autocorrelation function $R(\tau)$. Note that the period T_0 should be measured from the oscillation having a larger amplitude in the function $R(\tau)$, which means the range of a smaller τ because the positions of the oscillation maxima and minima are erroneously shifted in the range of rather attenuated oscillation or a larger τ. Unless the object's velocity is constant in the probing volume, the power spectrum of the periodic component may be asymmetrical, and the spectrum peak does not correspond to the central frequency f_0. In this case, the distance between the first maximum (at $\tau = 0$) and the second maximum in the correlation function does not give the true signal period. It is, therefore, advisable to measure distances between any other two successive maxima and to average them for more accurate measurements of f_0 [4].

If the periodic signal has a continuous random amplitude consisting of many bursts from the random passage of particles or from rough surfaces, the contributions from particles or scattering centers are uncorrelated with each other. Then, the autocorrelation function may be unchanged and used to determine the central frequency as well as above. Also note that the rate of

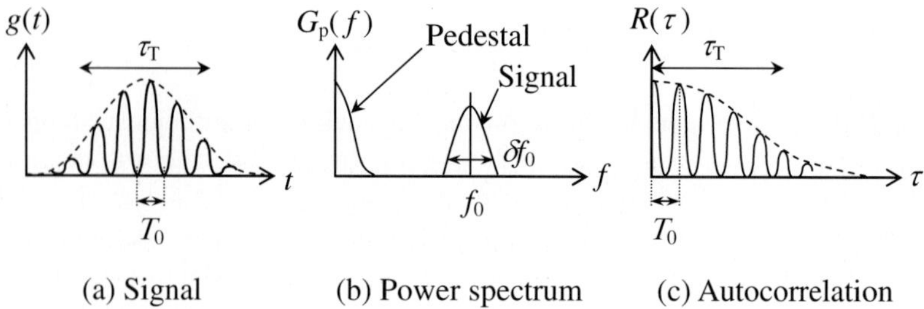

(a) Signal   (b) Power spectrum   (c) Autocorrelation

Fig. 4.7. (a) Typical burst-signal model with a Gaussian envelope, (b) the corresponding power spectrum, and (c) the corresponding autocorrelation function

attenuation in the correlation function, which can be specified by τ_T, is related to the peak width δf_0 of the periodic component in the power spectrum shown in Fig. 4.7b. By assuming Gaussian broadening for the signal spectrum, the width δf_0 becomes

$$\delta f_0 = \frac{2}{\pi \tau_T} . \tag{4.7}$$

If the oscillatory attenuation of the correlation function is slow or τ_T is relatively large, the signal period T_0 may be measured accurately from the distance between two successive maxima or minima. This corresponds to the fact that the central frequency f_0 can be determined more accurately from a narrower signal spectrum. Measuring the period T_0 of the autocorrelation function $R(\tau)$ might be considered equal to determining the period T_0 directly from the signal $g(t)$. However, this is not true because reading the period of the signal is substantially influenced by noise and resultant period data potentially involve errors. In the autocorrelation function, the effect of noise is considerably reduced by integration long compared to the signal period included in the correlation principle. Then, more accurate measurements of period T_0 can be made from the autocorrelation function than from the direct signal.

The instrument to calculate and display the correlation function of an analog input signal is an electronic correlator. However, the analog correlator is at present hardly employed for SFV signal processing systems. Recent correlators operate mostly in digital circuits to speed up the calculation and to improve the accuracy and the commercial cost. The input photocurrent, thus, must be sampled and converted to digital values by the analog-to-digital converter (ADC). The processing time is generally limited by the clock frequency of the ADC and the number of data accumulated in processing. Another approach [90, 91] is to perform the digital correlation analysis of zero crossings [53] obtained from the analog signal. Once the analog signal is digitized, the further analysis in this approach is the same as that in the digital operation in the framework of the photon correlation technique, which will be described in Sect. 4.5.3.

4.5.2 Fast Fourier Transform

In general, the power spectrum gives information necessary for velocity determination in SFV measurements, that is, the central frequency f_0 and, if required, its dispersion for evaluating the velocity fluctuation. If, however, the autocorrelation function is desired, it is another way to execute the Fourier transform of the power spectrum obtained by spectrum analysis described in Sect. 4.2 based on the Wiener–Khintchine theorem. For this process, the FFT algorithm can be used favorably. Commercially available FFT processors are usually equipped with this additional operation, and, thus, they can always display both the power spectrum and the correlation function simultaneously, obtained from one signal input. This approach is advantageous for shortening the processing time in comparison with direct correlation computing.

4.5.3 Photon Correlation Technique

When SFV measurements are performed at very low light levels, the continuous intensity signal cannot be detected due to its level below random photon noise. Even in this case, the photoelectron pulses carry useful information of the optical signal. These pulses appear in a discrete manner and can be processed about directly in digital electronics. The photon correlation technique deals with the discrete correlation analysis of counts of photoelectron pulses. Thus, it is particularly useful for low light levels. If the light level increases and exceeds a certain limit, individual photoelectron emissions are no longer isolated, and the result is a continuous intensity-fluctuating signal. There is, thus, a limit on the maximum acceptable light level. The photon correlation technique has been considerably well established [92], and commercial instruments are available for SFV as well as LDV measurements.

As shown in Fig. 4.8, at very low light levels, the probability of occurrence of photoelectron pulses or the number density of pulses at a given time is proportional to the light intensity (**a**). The input pulse train (**b**) obtained by the photon counting detector is sequentially counted within a prescribed sampling interval $\delta\tau$ (usually a few µs, for example). The pulse counts m_j ($j = 1$–N) (**c**) are, then, used to calculate the autocorrelation function (**d**) in a discrete manner. The discrete form of (4.6) is given by

$$R\left(q\delta\tau\right) = \sum_{j=1}^{N} m_j m_{j+q} \,, \tag{4.8}$$

where N is the total number of terms in the summation and q is an integer giving a delay time increment and usually corresponds to the channel number of the digital correlator (typically 64–256 channels). This means that the correlator executes q-channels of summations simultaneously . Detectors used for photon counting are in general photomultipliers specifically designed to

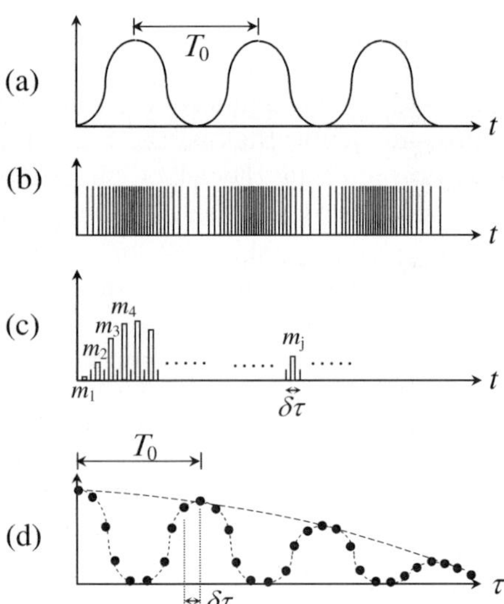

Fig. 4.8. Schematic illustration of the photon counting procedure for the photon correlation technique

suppress dark current noise. In consequence of recent advances in semiconductors, avalanche photodiodes (APDs) [93–96] may also be suitable devices for building compact photon correlation systems. The detector output contains signal pulses and dark current pulses. By analyzing these pulses with a pulse height analyzer, the resultant distribution [93] generally demonstrates that signal pulses are distributed in a larger range, whereas dark pulses are in a lower range of their heights. To discriminate signal pulses from dark pulses by using the criterion of pulse height, the photon counting unit has a pulse height discriminator, which sets a certain threshold for the pulse height to trigger signal pulse generation. The threshold level should be above the height of dark pulses but as low as possible to ensure full use of the signal pulses.

The implementation of (4.8) is entrusted to digital operations [97, 98] in the correlator and not described in this book. The main operation is the time-sequential multiplications of photon counts and their q-channel summations by using counters, shifting registers, and storage channels. To reduce operating time and to realize high-speed performance, practical correlators often employ a "clipping" technique [99], in which the pulse counts are classified into high (or "1") and low (or "0") levels by using a preset criterion or so-called "clipping level." After clipping, the processing is basically one-bit operation and much simplified. The measured photon correlation function (**d**) in Fig. 4.8 shows the same oscillatory form as Fig. 4.7c, and the period T_0 between the two successive maxima or minima is read to determine the central frequency f_0. If necessary, the determination of the frequency f_0 can also be made in the power spectrum obtained by the Fourier transform of the resultant photon correlation function. The photon correlation technique is definitely advantageous for a poor signal-to-noise ratio at low light levels, but it has a limitation for real time operation. Modern digital electronics is improving the circumstances for correlation analysis. Real time monitoring of the velocity variation at very low light levels may be realized by advanced photon correlators.

4.6 Choice of the Signal-Analyzing Technique

A variety of circumstances influence the signals in SFV measurements, and the types of signals also vary: continuous and intermittent, strong and weak in intensity, good and poor SNRs, and frequency-constant and variable. Applications often require the mean velocity and also the velocity fluctuation or distribution and their real time recording. Since, however, there is no ideal signal-analyzing technique that is suitable for all situations, a proper choice is the key step to direct successful SFV measurements.

Providing that the velocity is almost constant and that the signal is continuous with a fair SNR, almost any technique is available for processing. In general, the frequency tracker and the frequency counter may be reasonable choices. If the signal frequency is lower than MHz, the FFT processor is very convenient and economical. The spectrum analyzer of the frequency scanning type is good for general purposes if sufficient time is given for processing. The filter bank, wave-period measuring technique, and photon correlator usually require elaborate hardware and are costly. Thus, use of these techniques should be considered in the limited cases where any other technique is not available. For intermittent signals at constant velocity, counting techniques are recommended. If the signal is of poor quality, tracker and spectrum analyzers can be used and, for extremely low light intensity, the photon correlation technique may be an ultimate choice.

When the velocity fluctuation is of major significance, care has to be given to performance on the timescale. It is, of course, true that the frequency tracker is the best choice for continuous signals whose velocity fluctuates. For intermittent signals, the tracker is still available as long as the timescale for the velocity fluctuation is short compared with the mean period between signal bursts. For very poor intermittent signals, the tracker no longer works satisfactorily, and the frequency counter and the wave-period measuring technique may be alternate means. If the SNR is poor, the best choice can be the filter bank, although the frequency resolution is degraded. The FFT processor and the photon correlator may be possible for signals in the frequency range lower than MHz. In this case, the integration time should be shorter than the timescale of the velocity fluctuation. Otherwise, the result gives only a mean value for the velocity fluctuation.

There is an alternative method for measuring velocity fluctuation, which is described only briefly below. The method needs two simultaneous output signals whose central frequencies are equal to each other and whose phases are orthogonal. One of the signals is treated as the real part and the other as the imaginary part, and these outputs form an analytic signal. Instantaneous amplitudes of the two phase-orthogonal signals produce the instantaneous phase of the analytic signal and, thus, the time rate of the phase change gives the instantaneous values of the central frequency and the object's velocity. To employ this method, the SFV system is required to output a set of phase-orthogonal signals and, then, availability and implementation of the method depend on the spatial filtering device and system used. Therefore, the method will be described in Sect. 5.8 in relation to the introduction of imaging-type spatial filtering devices.

Spectrum analyzers of the frequency scanning and FFT types, and the photon correlator for general uses are commercially-available instruments. Some different types of frequency trackers and frequency counters are also provided by manufacturers exclusively for LDV measurements and can be used for SFV measurements. Probably, the filter bank processor and wave-period measuring system have to be designed to meet required specifications. Table 4.1 summarizes the performance of various processing techniques described above. The estimates listed assume normal uses of typical instruments in each technique and may be useful for rough and brief comparisons.

Table 4.1. General performance of processing techniques for SFV signals

Processing techniques		Velocity fluctuations	Intermittent signals	Low SNR	Freq. range	Accuracy	Temporal resolution
Spectrum analysis	Frequency scanning	×	⊙	⊙	∼ GHz	⊙	×
	Filter bank	⊙	⊙	⊙	∼ GHz	×	⊙
	FFT	△	⊙	⊙	∼ MHz	⊙	⊙
Frequency tracking	Frequency tracker	○	×	⊙	several tens of MHz	⊙	⊙
Counting technique	Frequency counter	⊙	⊙	×	∼ GHz	○	⊙
	Wave-period measurements	⊙	○	×	∼ MHz	○	○
Correlation analysis	Photon correlation	△	⊙	○	several tens of MHz	⊙	△

○ Excellent or very good; ⊙ good; △ fair or not bad; × difficult or unavailable

5

Spatial Filtering Devices and Systems

Various types of velocimeters based on the principle of the spatial filtering method, have been reported so far. Spatial filtering velocimeters are characterized mainly by both the spatial filtering device and the measuring system employed in each type of velocimeter. The signal quality depends primarily on the spatial filtering characteristics. System performance, including measuring functions, operations, and applicability, features various types of optical and signal-analyzing systems. As described in Chap. 2, the basic type of spatial filtering velocimeter is constructed from a transmission grating and an optical imaging system with a single lens. To improve the performance, expand the usefulness, and develop a practical system, a variety of spatial filtering devices have been studied with suitable optical and signal-analyzing systems. Those types may be roughly divided into eight categories.

Starting from the transmission grating as a basic type, this chapter presents the eight categories of spatial filtering devices and systems. In Sect. 5.1, the transmission grating type is described, including some derivative techniques to remove the pedestal component or the directional ambiguity and to measure two-dimensional velocity components. Sections 5.2–5.8 treat other advanced or specifically designed types: prism grating, lenticular grating, optical fiber array, liquid crystal cell array, integrated solar cell array, one-dimensional detector array, and two-dimensional image sensor. Section 5.9 summarizes these devices on the basis of measuring functions; pedestal removal, directional discrimination, and two-dimensional velocity measurements.

5.1 Transmission Grating

A transmission grating is a simple optical element consisting of a parallel-slit reticle with opaque and transparent bars. A typical example is the Ronchi grating (ruling), as shown in Fig. 5.1, which is generally used for investigating the shapes of optical surfaces, known as the Ronchi test [100]. This grating is

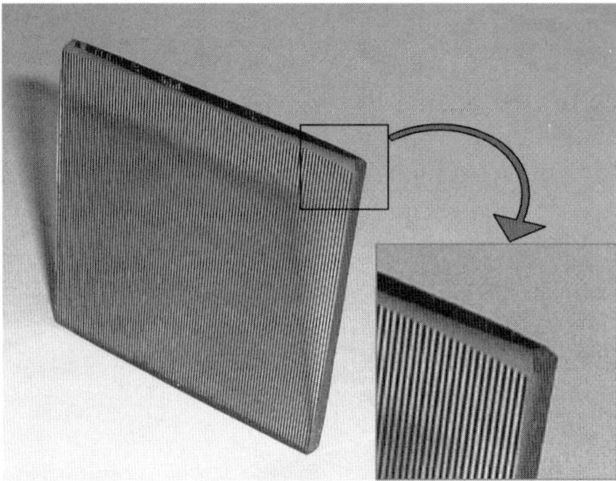

Fig. 5.1. A Ronchi grating (ruling) with a pitch of 508 µm

made by vacuum evaporation of chromium on a glass substrate, and the ruling of $50 \sim 500$ lines per inch is quite popular [101]. This range of ruling pitch is larger than that used for a diffraction grating and then causes no significant diffraction effect. Use of a transmission grating in the imaging plane realizes simply the most fundamental spatial filtering velocimeter.

5.1.1 Transmission Grating Velocimetry

The basic optical system of the velocimeter using a transmission grating is depicted in Fig. 5.2. The measuring principle was already described in Sect. 1.2. A transmission grating G is attached with a mask M to define a probe volume in the object plane x_0–y_0 and is placed in the image plane x–y to realize the spatial filter (SF). Light passing through the spatial filter is collected by a lens L_2 and received by a photodetector (PD). With the movement of objects, periodic signals are observed and analyzed to derive the velocity of moving objects according to the SFV principle (see Sects. 1.2 or 2.1). When the moving direction of objects is known and constant, the spatial filter is oriented so that its grating lines are normal to the moving direction. The optical imaging system is chosen flexibly to fit measurement circumstances and purposes, such as microscopic or telescopic imaging, and imaging with camera lenses, including zoom lenses and wide-angle lenses. If a commercially available Ronchi grating is to be used for the transmission grating, available grating pitches limited to a certain range around 50–500 µm typically. Then, adjustment of the imaging magnification gives freedom to design an optimum spatial filter. A photodiode is typically used to detect signals and, for a low light level, an avalanche photodiode or photomultiplier tube (PMT) may be useful. Output signals are processed by typical signal-analyzing methods such as spectrum analysis or frequency counting, as described in Chap. 4. When the scattered light intensity is extremely low, a photon-counting type PMT is an

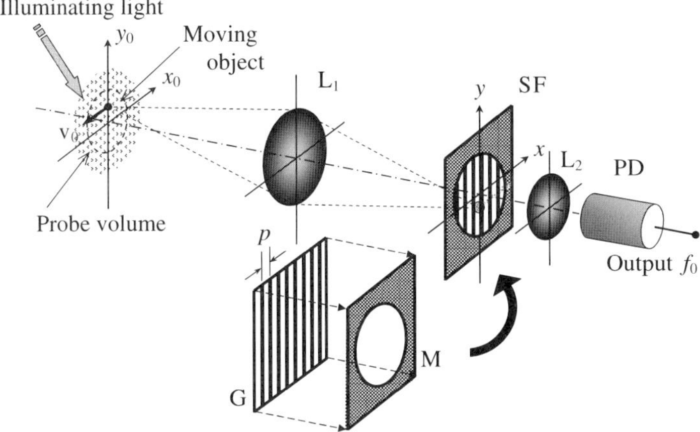

Fig. 5.2. Basic optical system of a transmission grating velocimeter

effective choice for signal detection in the framework of the photon correlation technique. The freedom of choice of an optical imaging system, transmission grating, and photodetector type is a very attractive feature for practical uses.

Since the transmission grating itself is simple and easy to handle, a compact and stable velocimeter can be constructed, that is flexible for measurements in various practical situations, especially when other complicated spatial filters are unavailable. Due to its basic typical feature, this type of velocimeter is often referred to as a transmission grating velocimeter [28]. Historically speaking, the transmission grating velocimeter has already been treated for application-oriented studies in the beginning stages of SFV development. In 1964, Gaster [14] measured a liquid flow velocity lower than 0.5 cm/s using a transmission grating on a rotating disk. This technique was then used to measure the flow velocity in various conditions, for instance, in a microscopic region [23].

5.1.2 Differential Detection for Pedestal Removal

Figure 5.3 illustrates a typical model of output signals from a photodetector (PD) and the corresponding power spectrum. The signal consists of a periodic component containing information about the object's velocity and a lower frequency component called the pedestal. The former component is used for velocity measurements, whereas the latter is useless. To determine the velocity, the central frequency f_0 of the periodic component should be measured accurately. The pedestal component decreases the signal-to-noise ratio and causes derivation of the central frequency of periodic signals as described in Sect. 2.4.2. The ambient light can be a source of the dc component, which also reduces the signal-to-noise ratio. When counting-type techniques are employed for signal processing, the pedestal component must be removed

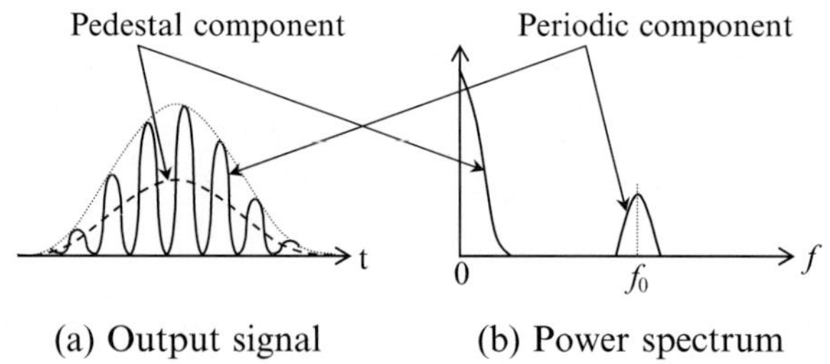

(a) Output signal   (b) Power spectrum

Fig. 5.3. Typical model of output signals and the corresponding power spectrum

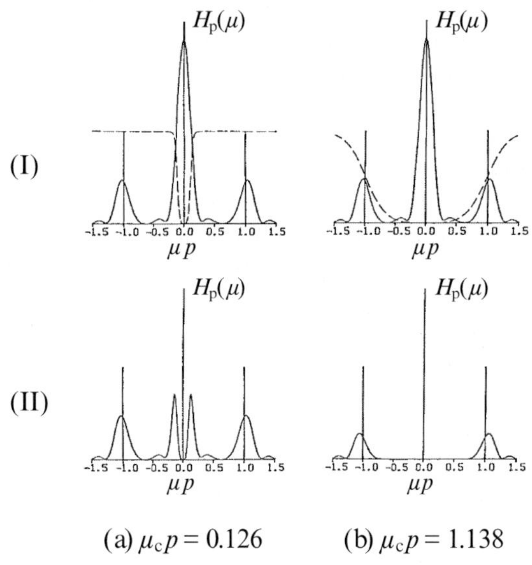

(a) $\mu_c p = 0.126$ (b) $\mu_c p = 1.138$

Fig. 5.4. (**I**) Computed power spectra of a spatial filter (*solid curves*) and an electric HPF (*dashed curves*) and (**II**) the corresponding spectra after filtering

from the photodetector signal before it is processed because the zero or level crossing is incorrectly counted if the pedestal is contained. Removing the pedestal is generally done by using an electric high-pass filter (HPF) since the pedestal consists of components with dc and lower frequencies than the central frequency f_0. The use of an electric HPF is a possible means if the periodic and pedestal components are sufficiently separated from each other in the spectral domain. This situation is realized when the number n of grating lines given by (2.37) is relatively large, such as $n \geq 10$. As shown in Fig. 2.12, decreasing number n of grating lines results in spectral broadening, and the resultant poor separations between the periodic and pedestal components make it difficult to remove the pedestal by using electric HPF.

Figure 5.4 shows the computed power spectra of a spatial filter (*solid curves*) and an electric HPF (*dashed curves*) in the upper part (**I**) and the

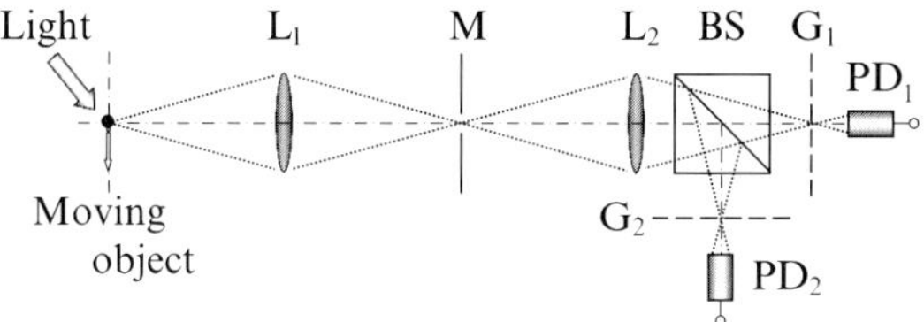

Fig. 5.5. Basic optical system of a transmission grating velocimeter using differential detection

corresponding spectra after filtering in the lower part (**II**). The spatial filter is assumed to be the circular type with sinusoidal transmittance of the grating number, $n = 4$. The frequency characteristics of the electric HPF are assumed to follow the Butterworth of the fourth order. The horizontal axis represents the spatial frequency μ that is normalized by $1/p$, when p is the grating interval. Then, $\mu p = 1$ corresponds to the central frequency f_0 of output signals in the temporal domain. In result (**a**), the normalized cutoff frequency $\mu_c p$ is set small enough to pass the periodic component, and, then, the filtered spectrum still contains a considerable pedestal component. Conversely, in (**b**), the cutoff frequency is increased to remove the pedestal sufficiently, and the HPF cuts a fraction of the periodic component, causing a serious deviation of the central (or peak) frequency. Thus, the use of an electric HPF is unsuitable for a small grating number n. Even if a value of n is not small, the periodic component spectrum is close to the pedestal when an object moves in a direction oblique to the grating lines, as described in Sect. 2.4.3. If the object's velocity is temporally variable, a choice of the cutoff frequency in the electric HPF is quite difficult. This is a source to limit the dynamic range of velocity measurements. In these circumstances, the pedestal should be removed in any other way.

A solution to this problem is the introduction of differential detection into the optical system. The basic method of differential detection employs two sets of spatial filters and photodetectors. Figure 5.5 schematically shows the basic optical system of a transmission grating velocimeter using differential detection. Two identical images of a moving object are formed on two transmission gratings G_1 and G_2 via lenses L_1 and L_2 and a beam splitter (BS). By placing a mask M in the first imaging plane, it is imaged with a lens L_2 on two gratings. This imaging gives completely identical windows to the two gratings and guarantees definition of the same probe volume to two photodetectors PD_1 and PD_2. Two output signals from the photodetectors are fed into the positive and negative inputs of a differential amplifier (DA). The two gratings G_1 and G_2 are placed so that their grating lines are parallel in an out-of-phase manner, as shown in Fig. 5.6. Then the differential amplifier cancels out the pedestal component, including the dc component due to the ambient light, and at the same time doubles the amplitude of the periodic component. Since the pedestal cancellation is made independently of the central frequency f_0 of periodic signals, a wide dynamic range of velocity can be obtained.

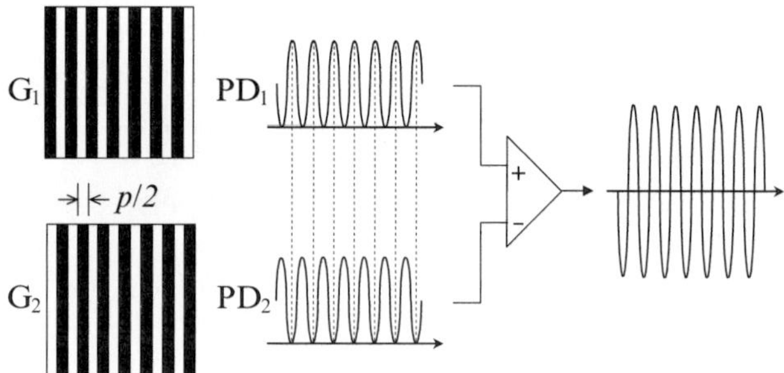

Fig. 5.6. Two gratings placed in an out-of-phase manner for the differential detection

The equivalent differential transmittance $h(x,y)$ formed by using the above-mentioned arrangement is expressed for the circular-type spatial filter with sinusoidal transmittance by

$$h\left(x,y\right) = h_1\left(x,y\right) - h_2\left(x,y\right) \,, \tag{5.1}$$

where $h_1(x,y)$ and $h_2(x,y)$ are transmittance functions of the two gratings G_1 and G_2 with circular window images having a radius of a, respectively, and are written as

$$h_1\left(x,y\right) = \begin{cases} \dfrac{1}{4}\left(1 + \cos\dfrac{2\pi}{p}x\right), & x^2 + y^2 \leq a^2 \,, \\ 0\,, & \text{otherwise,} \end{cases} \tag{5.2}$$

$$h_2\left(x,y\right) = \begin{cases} \dfrac{1}{4}\left[1 + \cos\left(\dfrac{2\pi}{p}x - \pi\right)\right], & x^2 + y^2 \leq a^2 \,, \\ 0\,, & \text{otherwise.} \end{cases} \tag{5.3}$$

A factor of $1/4$ on the right-hand side instead of $1/2$ in (2.29) means the intensity reduction by half due to the use of the beam splitter. The corresponding power spectrum is derived by performing the integration of (2.5) with (5.1) as

$$\left|H\left(\mu,\nu\right)\right|^2 = \frac{\pi^2 a^4}{4}\left[H_J^-\left(\mu,\nu\right) + H_J^+\left(\mu,\nu\right)\right]^2 \,, \tag{5.4}$$

where $H_J^-\left(\mu,\nu\right)$ and $H_J^+\left(\mu,\nu\right)$ are the same as in (2.30). Comparison of (2.30) and (5.4) indicates that the pedestal component having a peak at $(\mu,\nu) = (0,0)$ disappears and that only the two periodic components having peaks at $(\mu,\nu) = (\pm 1/p, 0)$ are obtained in differential detection.

Figure 5.7 [25] presents power spectra $H_p(\mu)$ for four different numbers n of grating lines, obtained by integrating (5.4) with respect to spatial frequency ν. Comparison of Fig. 5.7 with Fig. 2.12 indicates that, in theory, the pedestal component disappears completely and that the central frequency or the peak of the periodic component appears correctly at $\mu p = \pm 1$ by using differential detection, even for small numbers of $n = 2$ and 5. Differential detection theoretically provides complete cancellation of the pedestal. To realize

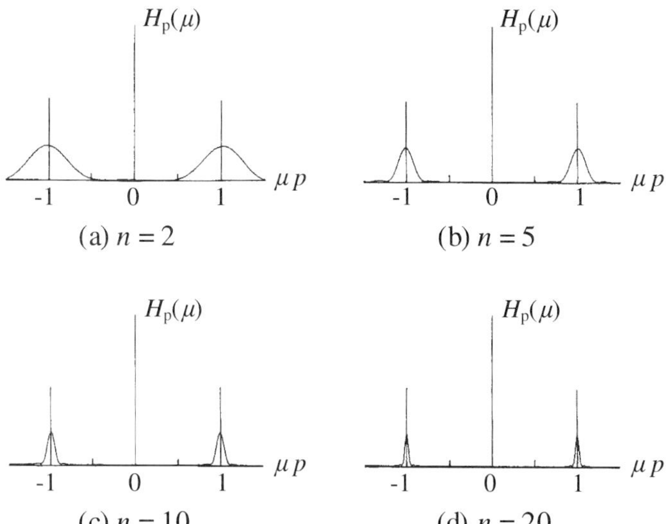

(a) $n = 2$ (b) $n = 5$

(c) $n = 10$ (d) $n = 20$

Fig. 5.7. Computed power spectra of the spatial filter in differential detection [25]

this operation actually, the out-of-phase positioning of two transmission gratings should be exact. A balance of sensitivity is also required between the two photodetectors. An error in these conditions may yield undesirable residual components and higher order harmonic noise in the differential output. To avoid this problem, an easier way [25] has been studied using a single spatial filter. This is simply realized by removing the transmission grating G_2 from the optical system of Fig. 5.5. In this alternative arrangement, photodetector PD_2 yields only the pedestal component as shown by the dashed curve in Fig. 5.3, due to the lack of grating lines. When the sensitivity of PD_2 is lowered to the half that of PD_1, the pedestal component is canceled out in the differential output, although the periodic component is not doubled in amplitude. This type of differential detection is, thus, advantageous for constructing a practical transmission-grating-type SFV system, which may be used for measurements of turbulent-flow velocity and the velocities in a microscopic region, for example. Figure 5.8 [25] demonstrates typical output signals and corresponding frequency histograms obtained by the wave-period measuring technique described in Sect. 4.4.2 for $n = 10$. Results (**a**) and (**b**) were obtained with the conventional-type arrangement of Fig. 5.2, whereas (**c**) was with the differential type using a single transmission grating. The pedestal component is reduced by an electric HPF with the cutoff frequency $f_c = 700\,\text{Hz}$ in Fig. 5.8a, whereas it is not removed in Fig. 5.8b due to $f_c = 50\,\text{Hz}$. In Fig. 5.8c, the pedestal is almost canceled out although an electric HPF of $f_c = 50\,\text{Hz}$ was used to suppress residual components. Comparison of Figs. 5.8a and c indicates that differential detection provides better results than the conventional type does since a narrower peak was obtained in frequency histogram (**c**) than in (**a**).

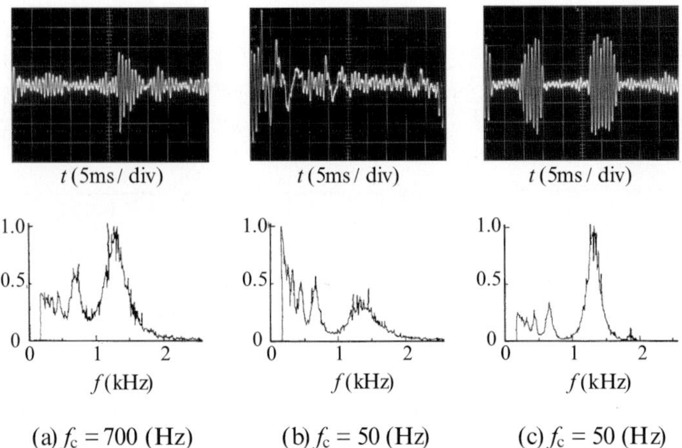

Fig. 5.8. Typical output signals and corresponding frequency histograms obtained by measurements with $n = 10$

5.1.3 Directional Discrimination — Frequency Shifting

A fundamental problem in SFV measurements is no sense of a velocity sign or direction. A change in the sign of an object's velocity gives no difference in frequency of output signals, and it is not possible to tell in the which direction the object is moving. This yields directional ambiguity, which is a common defect with other velocimeters such as LDV. The defect may be insignificant for some applications in which a moving direction is known or observable. However, changes in the flow direction are involved in many flow situations such as turbulent and recirculating flows and, thus, directional discrimination should be made in SFV measurements.

Several methods which are available for this purpose are based mainly on frequency shifting and phase shifting. The most popular methods involve shifting the central frequency of SFV output signals. The frequency shifting technique is commonly used for the same purpose in LDV where a frequency modulator or shifter is used. A diffraction grating or an acousto-optic Bragg

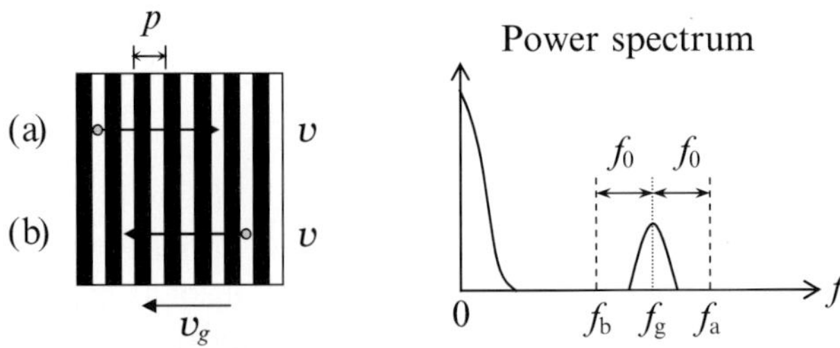

Fig. 5.9. Detection of an image motion with the moving transmission grating

cell produces a moving system of interference fringes in the differential LDV arrangement [5]. A similar effect is realized in SFV measurements by moving a grating in its own plane. When a transmission grating of pitch p is moving with velocity v_g as shown in Fig. 5.9, a stationary image on the grating generates a periodic signal of frequency $f_g = v_g/p$, which is the shift frequency due to the grating movement. Movement (**a**) of the image with velocity v in the direction opposite to the grating raises the signal frequency to $f_a = f_g + f_0$, ($f_0 = v/p$), whereas movement (**b**) in the same direction lowers it to $f_b = f_g - f_0$. Thus, the direction of the velocity is determined. To ensure that the lowered frequency f_b does not change sign, the frequency shift f_g must be larger than the original signal frequency f_0 corresponding to the maximum velocity expected in measurements. This condition maintains an unambiguous relation between the signal frequency and the velocity to be measured.

A mathematical expression for frequency shifting technique is briefly given below for a circular-type transmission grating with sinusoidal transmittance (see Fig. 2.10IIa). By assuming that the grating is moving with velocity v_g in the x direction, the transmittance function $h(x, y)$ is written, instead of (2.29), as

$$
h(x, y) = \begin{cases} \dfrac{1}{2}\left[1 + \cos\dfrac{2\pi}{p}(x - v_g t)\right], & x^2 + y^2 \leq a^2, \\ 0, & \text{otherwise.} \end{cases} \tag{5.5}
$$

Substitution of (5.5) in (2.5) yields the power spectrum of the spatial filter,

$$
|H(\mu, \nu)|^2 = \pi^2 a^4 \left\{ H_J(\mu, \nu) + \frac{1}{2}\left[H_{\mathrm{Jk}}^-(\mu, \nu) + H_{\mathrm{Jk}}^+(\mu, \nu)\right] \right\}^2, \tag{5.6}
$$

where $H_J(\mu, \nu)$ is the same as used in (2.30), and

$$
H_{\mathrm{Jk}}^-(\mu, \nu) = \frac{J_1\left[2\pi a\sqrt{\left(\mu - \dfrac{k}{p}\right)^2 + \nu^2}\right]}{2\pi a\sqrt{\left(\mu - \dfrac{k}{p}\right)^2 + \nu^2}},
$$

$$
H_{\mathrm{Jk}}^+(\mu, \nu) = \frac{J_1\left[2\pi a\sqrt{\left(\mu + \dfrac{k}{p}\right)^2 + \nu^2}\right]}{2\pi a\sqrt{\left(\mu + \dfrac{k}{p}\right)^2 + \nu^2}},
$$

where $k = 1 - (v_g t/x) = 1 - (v_g/v)$ when the image is assumed to move in the x direction with velocity v. Equation (5.6) indicates that periodic signal components have peaks at $(\mu, \nu) = (\pm[1 - (v_g/v)]/p, 0)$. According to the discussion of converting a spatial-domain treatment to a temporal one as described in Sect. 2.1, the frequency of periodic signal components that is actually observed is given, with the condition $f_{\mathrm{g}} > f_0$, by

$$f = \mu v = \frac{v_{\mathrm{g}} - v}{p} = f_{\mathrm{g}} - f_0 = f_{\mathrm{b}} . \tag{5.7}$$

An opposite sign of the image velocity v to the grating velocity v_{g} makes $f_{\mathrm{g}} + f_0 = f_{\mathrm{a}}$, thus shifting the frequency up.

A simple means for moving grating lines is using the rotation of a disk grating [14] or cylinder grating, as shown in Fig. 5.10. When a moving image is formed on a specific circumferential area of radial grating lines in a rotating disk, which is specified by radius r, the velocity v_{g} of the grating lines in the imaged area is given by $2\pi r N$, where N is the number of rotations per second. Then, the frequency shift f_{g} results in $2\pi r N/p$. A large value of radius r gives a good approximation of rotating radial lines to translational parallel lines. If a cylinder grating is considered, a photodetector unit may be installed inside it. Movement of lines on a curved surface can be approximated well to in-plane movement with a large cylinder radius. Generally, the cylinder type is more convenient for constructing a compact and stable system than the disk.

Output signals can be processed in the same way as ordinary SFV signals. For example, spectral analysis yields the result $f_{\mathrm{g}} \pm f_0$, which is compared with the shift frequency f_{g} to recognize the direction of the object's velocity. The shift in frequency of the output signals means that the signal spectrum moves away from the pedestal in the frequency spectral domain. Thus, it is also advantageous for removing the pedestal with an electric HPF. For real time monitoring of the direction, a frequency discriminator is more advantageous, in which the central frequency for reference is tuned to the shift frequency f_{g}. Then, the discriminator output in voltage is linearly proportional to the frequency of the applied signal. When an object is stationary, the photodetector output contains frequency f_{g} and the discriminator yields a zero output.

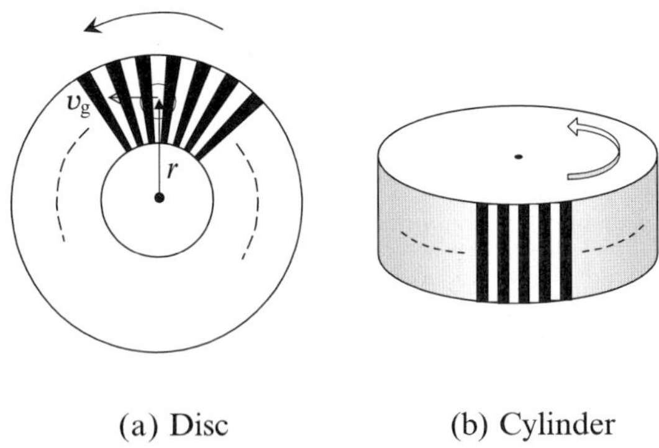

(a) Disc   (b) Cylinder

Fig. 5.10. (a)Rotating disk and (b)cylinder gratings

A higher or lower frequency than f_g is responded to by a positive or negative voltage in the discriminator output.

The frequency shifting technique is a simple and direct solution to directional discrimination, but realizing it with a rotating grating requires careful treatment. The system needs mechanical operation, which is a source of instability in output signals such as frequency noise due to the fluctuation of rotation. The rotating grating is also disadvantageous for producing a large frequency shift. This technique is therefore useful for relatively low velocity.

5.1.4 Directional Discrimination — Phase Shifting

Another method of determining whether an object's velocity is positive or negative is to detect SFV periodic signals in two channels having a phase difference of $\pi/2$ with each other and, then, to see which channel leads onto the other [26]. An optical system which realizes this idea is the same as Fig. 5.5, but two transmission gratings G_1 and G_2 for this purpose are placed in a phase difference $\pi/2$ or $p/4$, as shown in Fig. 5.11. Two output signals from photodetectors PD_1 and PD_2 generate the phase difference of $\pi/2$ (as shown also in Fig. 5.11), whose sign is determined by the direction of backward or forward image movement, according to a predetermined convention. Therefore, the velocity direction is revealed by detecting the sign of the phase difference. For this system to work, it is important that each photodetector receives exactly identical images in the same probe volume. Since this technique contains no mechanically moving part, a stable system can be constructed, although the initial arrangement of two gratings with the required phase difference and an appropriate signal processing (described below) is necessary.

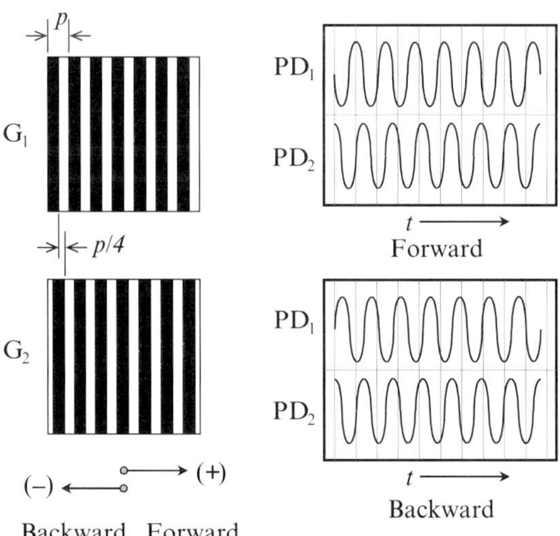

Fig. 5.11. Phase shifting technique for directional discrimination

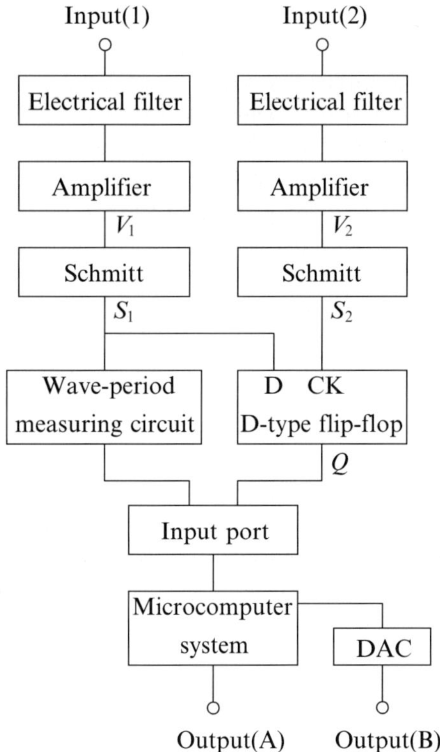

Fig. 5.12. Block diagram of signal processing for directional discrimination [26]

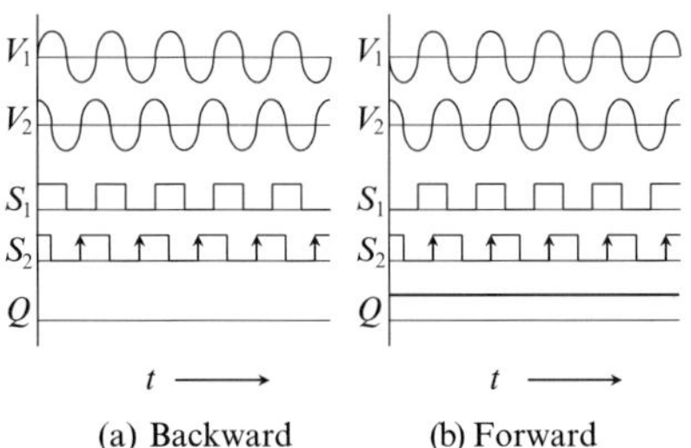

Fig. 5.13. Timing chart of signal processing for directional discrimination [26]

Figure 5.12 [26] shows an example of a signal-analyzing block diagram employed for the phase shifting technique. Two photodetector signals are fed, respectively, into Input (1) and Input (2). Outputs V_1 and V_2 obtained via electric filters and amplifiers are applied to Schmitt circuits by which square waves S_1 and S_2 are produced, as shown in the timing chart of Fig. 5.13 [26]. The signal S_1 is then sent to the wave-period measuring circuit (see Sect. 4.4.2) by which the central frequency f_0 is determined and used to derive the object's velocity in the microcomputer. To detect the sign of the phase difference between two input signals, output S_1 is also fed into a data input (D) of a D-type flip-flop (D-FF) circuit, whereas output S_2 is into a clock input (CK). In Fig. 5.13a, signal S_2 leads onto signal S_1 by $\pi/2$, which corresponds to the image moving backward in Fig. 5.11. Now, if the moving direction changes forward, the timing relation between the two photodetector signals turns over and, thus, S_1 leads onto S_2, as shown in Fig. 5.13b. The D-FF is used to determine this timing relation. In the example, the D-FF is triggered at every positive edge of square wave S_2 to output the logic state of square wave S_1. Thus, output Q of the D-FF is in logic state 0 (showing low level) for Fig. 5.13a, whereas it is in state 1 (high level) for Fig. 5.13b. The output Q is sent to a microcomputer by which the sign of the velocity is determined by examining the level of Q and is added onto the corresponding data of central frequency f_0 that is obtained from the wave-period measuring circuit. In this way, the system finally provides a negative or positive central frequency, which corresponds to the velocity in the backward or forward direction. Figures 5.14a and b demonstrate measured histograms of the central frequency f_0 obtained for image movement in the backward $(-)$ and forward $(+)$ directions, according to the convention of Fig. 5.11, respectively. Most of the frequency data for the backward direction appear in the negative frequency region, whereas those for the forward one are in the positive region. The magnitude and directions of the object's velocity can therefore be simultaneously determined by this processing.

In other ways, both tasks of measuring the central frequency and sensing the moving direction are combined in a single procedure. In the framework of the complex Fourier transform [102], the two periodic outputs V_1 and V_2 $\pi/2$ out of phase are treated as the complex (real and imaginary) input or the analytic signal in a complex Fourier spectrum analyzer from which the velocity as well as the direction of the object are obtained. Also, the instantaneous amplitudes of the two phase-orthogonal outputs V_1 and V_2 are used directly to obtain the instantaneous phase of the analytic signal. The

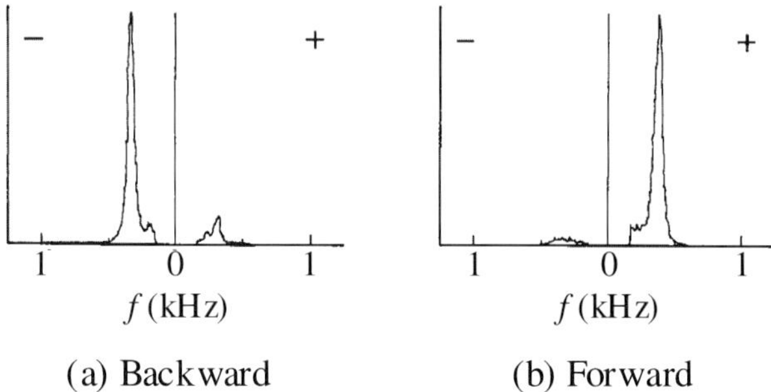

Fig. 5.14. Measured frequency histograms for images moving in backward and forward directions [26]

temporal changing rate or time differential rate of this phase gives us the instantaneous frequency and thus the velocity. Since the rotational direction or the increasing/decreasing of that phase is reflected in the sign of the instantaneous frequency, this is used for discriminating the moving direction. This latter method will be described in Sect. 5.8.

5.1.5 Two-Dimensional Measurements

In the basic arrangement of transmission grating velocimetry, as shown in Fig. 5.2, the velocity component perpendicular to the grating lines can be measured. Any other component may be observed by suitable rotation of the transmission grating in its own plane, but only one component is measured at a time. If two velocity components are measured successively, correlations between the two components being recorded may be degraded. A further interesting variation of the basic transmission grating velocimeter is, thus, to facilitate simultaneous measurements of two orthogonal velocity components.

An example of two-component measurements is introduced by using the optical system shown again in Fig. 5.5, but with a change in aligning the two gratings, that is an orthogonal orientation of the grating lines, as shown in Fig. 5.15. In this configuration, the central frequencies f_x and f_y of the two detector outputs are proportional to the two velocity components v_x and v_y, respectively. The real magnitude and orientation of the object's velocity are obtained trigonometrically by using the two components. If the two signals are known to contain different frequencies, the two output signals from the detectors are processed, in principle, by using a single-channel analyzing system such as an ordinary spectrum analyzer to determine the central frequency. When the two frequencies are close to each other, they are not separated in the spectral domain. Thus, practical uses require two channels of the signal-

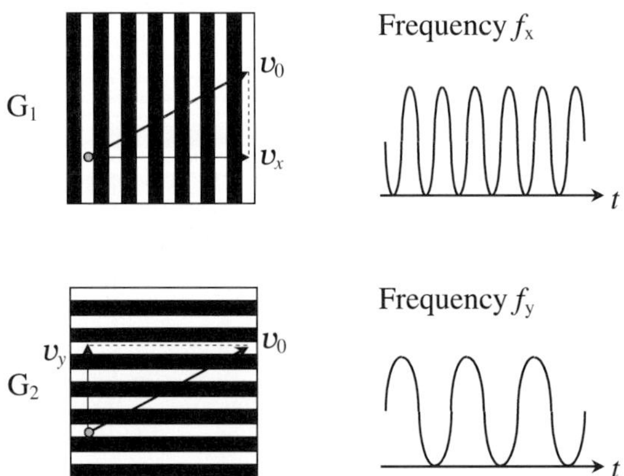

Fig. 5.15. Orthogonal alignment of grating lines and two output signals for two-dimensional measurements

analyzing system. As described in Sect. 5.1.2, the mask M in Fig. 5.5 is used to define the identical probe-cross sectional area for detectors PD_1 and PD_2. This imaging system ensures measurements of two velocity components in the same probe volume.

The measurement accuracy of each component v_x or v_y is highest in the moving direction perpendicular to the lines of each grating, and in other directions, it drops with an increase of the angle with respect to the direction normal to the grating lines. Since grating lines of G_1 and G_2 are orthogonal to each other, the measurable range of direction is limited to $\pi/2$. Velocities in other directions entail a change of sign, and true values will be determined with directional discrimination. Without using discrimination, improvement of the accuracy and the range of direction may be possible to some extent by using three or more transmission gratings. Figure 5.16 [103] illustrates an example of three gratings placed in a triangle. A moving image is detected by three photodetectors behind the gratings and the velocity and direction can be determined, in principle, from any two of three outputs. However, the study has shown that an estimation of the velocity and direction can be made more accurately by a weighted summation of the three outputs under the criterion of the least mean square error. This configuration enlarges the measurable range of direction to $2\pi/3$.

In the beginning study of the SFV by Ator [12], two-component measurement was discussed for correcting the drift angle of an aircraft. By using two gratings (reticles) in a herringbone pattern, as shown in Fig. 5.17 [12], two frequency outputs are compared to detect the drift angle. A servomechanism can be used to position the physical heading of the air plane automatically to the

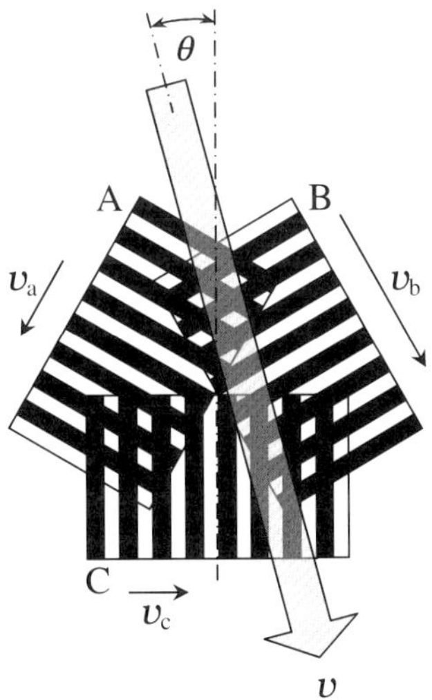

Fig. 5.16. Use of three transmission gratings for two-dimensional measurements (re-drawn from [103])

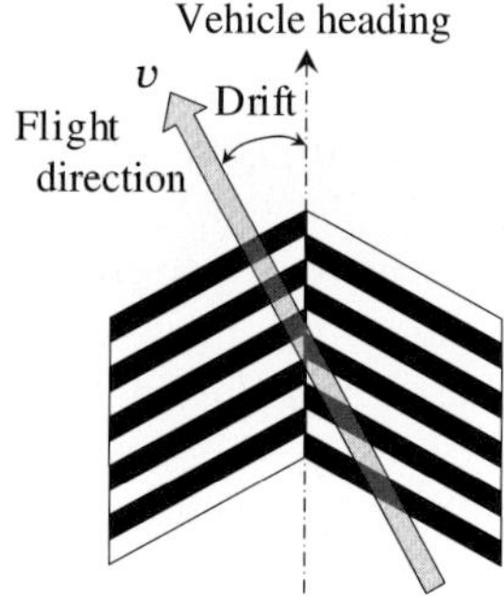

Fig. 5.17. Herringbone alignment of grating lines [12]

true line of flight by maintaining a zero-frequency difference. This technique is used for directional control of various moving objects.

To determinate the velocity direction of a vector from two-component measurements, it is necessary to perform directional discrimination for the two components and to detect changes in sign. This is realized by introducing the frequency shifting technique described in Sect. 5.1.3 to the two detection channels. The phase shifting technique is unsuitable for practical measurement systems since one-component measurement requires two detection channels and two requires four. The frequency shifting in two-component measurements is studied [104–106] with a single rotating disk having two transmission-grating rings, as shown in Fig. 5.18 [106]. The two gratings are formed annularly with different radii in the circumference of the disk, and their grating lines are ruled perpendicularly to each other ($\pm\pi/4$ with respect to the x axis). Imaging regions G_x and G_y are illustrated by circles on the two gratings and also depicted in the magnified scale. Two axes x' and y' are defined with an angle of $\pi/4$, respectively, against axes x and y. The inner and outer gratings have the same line interval p. Lines in each imaging region may be assumed to be parallel if the radii of two grating rings are much larger than those of imaging regions. Two identical images of a moving object are formed in the two grating regions G_x and G_y on the rotating disk. When the disk is rotated clockwise, grating lines of G_x and G_y move toward positive x' and negative y', producing shift frequencies $f_{gx'}$ and $f_{gy'}$, $(f_{gy'} > f_{gx'} > 0)$, respectively. Then, the central frequencies $f_{x'}$ and $f_{y'}$, $(f_{x'} > 0, f_{y'} > 0)$ of the two periodic signals from the photodetectors are given by the following equations:

$$f_{x'} = f_{gx'} - \frac{v_{ix'}}{p} , \tag{5.8}$$

$$f_{y'} = f_{gy'} + \frac{v_{iy'}}{p} , \tag{5.9}$$

where $v_{ix'}$ and $v_{iy'}$ denote the x' and y' components of the image velocity v_i. By measuring $f_{x'}$ and $f_{y'}$, $v_{ix'}$ and $v_{iy'}$ can be determined from (5.8) and (5.9) with known values of $f_{gx'}$, $f_{gy'}$ and p. The signs of $v_{ix'}$ and $v_{iy'}$ are determined

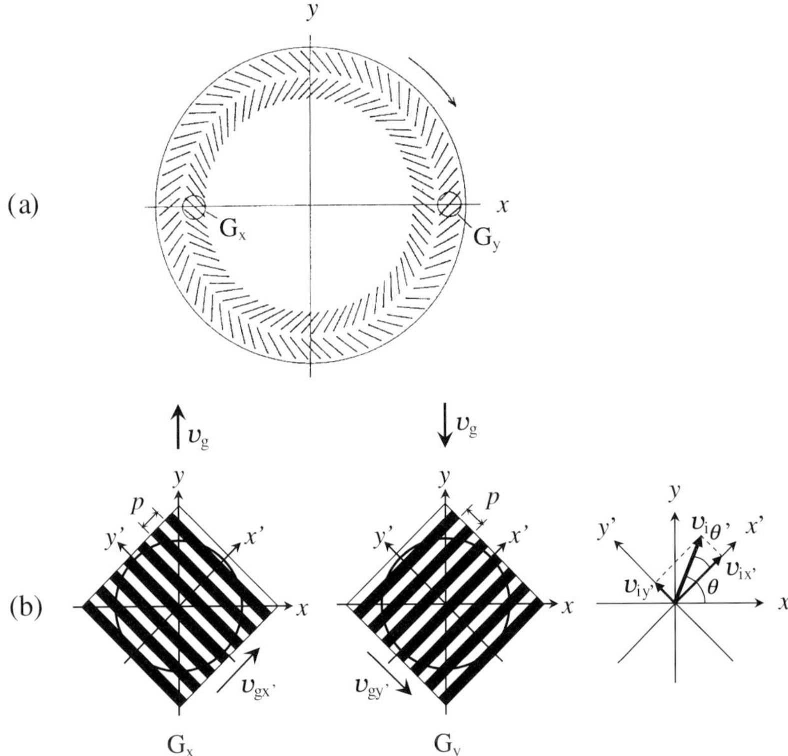

Fig. 5.18. Use of a single rotating disk having two transmission-grating rings for two-component measurements [106]

in accordance with the values of $f_{x'}$ and $f_{gx'}$, and $f_{y'}$ and $f_{gy'}$, respectively. For example, $f_{gx'} > f_{x'}$ gives $v_{ix'} > 0$ and $f_{gy'} < f_{y'}$ gives $v_{iy'} > 0$, thus the angle θ' of the image velocity v_i with respect to the x' axis is in the range of $(0, \pi/2)$ in this case. The direction of the image velocity v_i in (x, y) coordinates is expressed by the polar angle θ with respect to the x axis as follows:

$$\theta = \tan^{-1}\left(\frac{v_{iy'}}{v_{ix'}}\right) + \frac{\pi}{4} . \tag{5.10}$$

The real magnitude of v_i is given by the absolute value as

$$|v_i| = \sqrt{v_{ix'}{}^2 + v_{iy'}{}^2} . \tag{5.11}$$

As described in Sect. 5.1.3, the frequency shifting technique usually requires the grating velocity to be larger than the expected maximum velocity of a moving object's image. Then, the maximum grating velocity available in a measuring system determines the measurable velocity range. The grating velocity may be controlled by the rotating speed and the size of the disk and, thus, there is flexibility for varying the velocity range. The measurement accuracy in this example may be determined by the precision of both the ruling of the grating lines and the rotation of the disk.

5.2 Prism Grating

Light emerging from a prism is deflected in a direction different from that of the incidence, unless the input and output faces of the prism are parallel. This deflection effect can be used for the spatial filtering technique if it occurs in a spatially periodic manner. A device which is available for this purpose is a prism grating. There are two kinds of prism gratings which have been reported as spatial filtering devices: two-stage [40, 41] and three-stage [38, 39] types. Another device which is categorized as a prism type is the mirror grating [37]. No study has been reported in detail for investigating this device and, thus, it is only conceptually introduced in this book.

5.2.1 Two-Stage Type

Figure 5.19 shows schematically a two-stage prism grating and its use as a spatial filtering velocimeter [41]. An image of a moving object is formed on a flat face of the prism grating (PG). Light emerging from the prism is deflected in two different directions according to Snell's law at two facets consisting of periodic triangular ridges and, then, collected and received by photodetectors PD_1 and PD_2. Due to the movement of the image on the prism face, the prism grating deflects the emerging light alternately in the two directions. The two detector outputs result in periodic signals with a phase difference of π, which

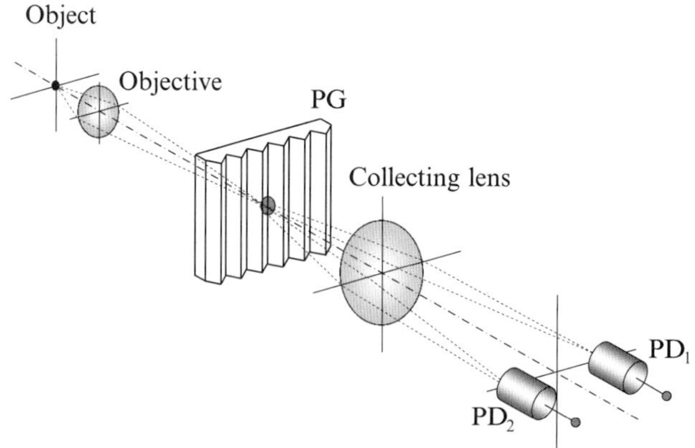

Fig. 5.19. Spatial filtering velocimeter using a two-stage prism grating [41]

are equivalent to the signals of Fig. 5.6 obtained by using two transmission gratings via a beam splitter, as shown in Fig. 5.5. Using a differential amplifier, the two outputs are differentially detected to enhance the signal modulation and to remove the pedestal component together with other dc. noise such as the ambient light.

In this way, a two-stage prism grating acts as two transmission gratings and also as a beam splitter. The prism grating also has, in principle, no significant loss of light due to opaque bars as in the use of a transmission grating. In these respects, the prism grating is a suitable device for constructing SFV systems. This grating can be introduced into various types of optical imaging systems together with a variety of photodetectors, such as a microscope with detection by a photomultiplier to measure blood flow velocity in microvessels. Note, however, that a prism has wavelength dispersion. When white light is used for illumination, the deflection angle of the light contains ranges, and therefore, collecting lenses and photodetectors should be positioned carefully. A laser source may be suitable for this device. It is probably difficult to obtain various sizes and grating pitches of the prism device, since usually, limited ranges are supplied by manufacturers. Pitches of $200 \sim 400\,\mu\mathrm{m}$ are typical examples [39–41].

5.2.2 Three-Stage Type

Although the two-stage prism grating is effective for removing the pedestal component, the SFV system using this device is unable to resolve the directional ambiguity. In transmission grating velocimetry, both the phase-shifting and frequency-shifting techniques are available for discriminating direction. In using the prism grating, however, frequency shifting by mechanical rotation is impractical. To realize directional discrimination by using the phase-shifting technique, a three-stage type of the prism grating has been developed. Figure 5.20 [39] illustrates the basic principle of the SFV system using a three-stage prism grating. As the object image moves on the flat face of the prism grating, light passing through the grating is repeatedly deflected in three different directions in consecutive order due to the refraction of light at three facets with different tilts. Such deflected light rays are received in three shifts by photodetectors PD_1, PD_2, and PD_3. For a shift of the deflected ray from one photodetector to the next, a displacement of $p/3$ is required for the image on the prism grating. Then, the time shift among the three photodetector outputs is one-third of a cycle, the phase difference $2\pi/3$. The central frequencies of the three signals are the same, and each of them is used to determine the object's velocity. Their phase relation reveals the direction of movement, for example, since signal 1 precedes signal 2 and signal 2 precedes signal 3.

Direct outputs of the three photodetectors contain the pedestal component and it should be eliminated. For this purpose, a method [39] using symmetry at linear transformation is applied, in which three outputs $S_1(t)$, $S_2(t)$, and $S_3(t)$ are appropriately added and subtracted with different weight factors a,

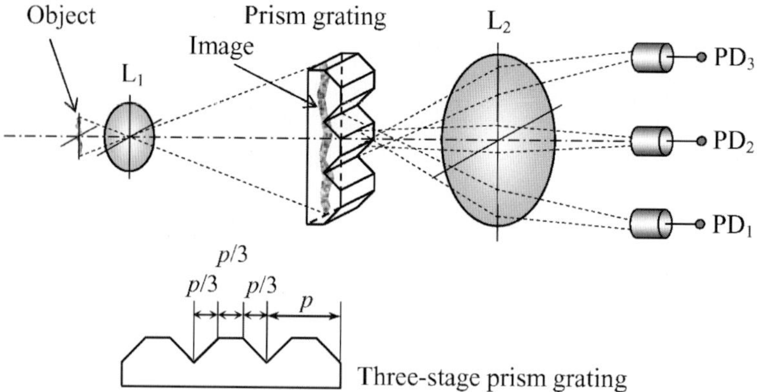

Fig. 5.20. Spatial filtering velocimeter using a three-stage prism grating (re-drawn from [39], Copyright 1981, with permission from Elsevier)

b, and c to generate two signals $A(t)$ and $B(t)$ with a phase difference of $\pi/2$. This processing can be designed so that $A(t)$ and $B(t)$ contain no pedestal component but keep the original central frequency. The weighting factors are derived by using the following relations;

$$A(t) = aS_1(t) + bS_2(t) + cS_3(t) , \tag{5.12}$$

$$B(t) = cS_1(t) + bS_2(t) + aS_3(t) , \tag{5.13}$$

$$a + b + c = 0 , \tag{5.14}$$

where $A(t)$ is perpendicular to $B(t)$ in phase angle. When $S_1(t)$, $S_2(t)$, and $S_3(t)$ are assumed to be sine functions with equal amplitude, the above transformation is easily realized, giving

$$\begin{cases} a = \left(-\sqrt{2} + \sqrt{6}\right)/6 , \\ b = \sqrt{2}/3 , \\ c = -\left(\sqrt{2} + \sqrt{6}\right)/6 . \end{cases} \tag{5.15}$$

The sign of the $\pi/2$ phase difference is assessed by appropriate signal processing, such as Fig. 5.12 in Sect. 5.1.4, to determine the direction of movement. Also, the phase-orthogonal signal pair $A(t)$ and $B(t)$ is used to derive the instantaneous frequency and velocity, whose sign gives the moving direction, in terms of the analytic signal. This will be treated in Sect. 5.8.

For each of the three photodetectors, the three-stage prism grating acts as a spatial filter having a slit width $w = p/3$ (p is the grating pitch, see Fig. 2.3 in Sect. 2.2). As can be seen from plots of (2.21) in Fig. 2.8, the power spectrum for $w = p/3$ becomes zero at the third-order harmonic, $3/p$. Then, all the third-order frequencies are absent in the three photodetector outputs. This means that the second harmonic component is still important, which could be absent if a spatial filter of $w = p/2$ is employed. Since the fundamental central frequency is used to determine the object's velocity, a second

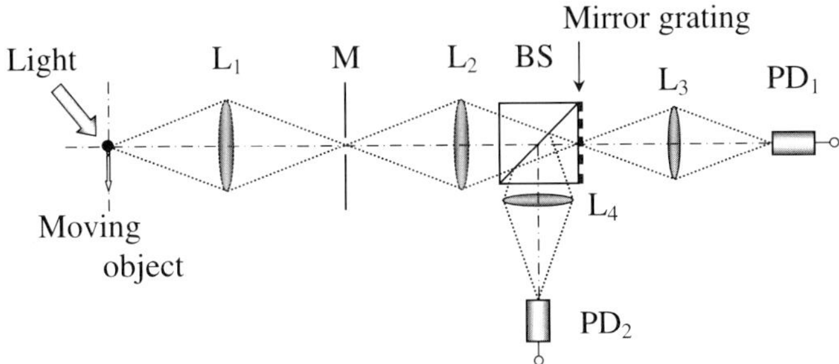

Fig. 5.21. Spatial filtering velocimeter using a mirror grating

harmonic next to the fundamental in the spectral domain is an undesirable component. If the number of grating lines, n, is small, the spectral bandwidth broadens and the second harmonic spectrum may have an influence on the fundamental frequency spectrum. This defect can cause the peak frequency to deviate, producing errors in a measured velocity. For a three-stage prism grating, therefore, a larger number of grating lines, at least more than 10, is desired.

5.2.3 Mirror Grating

If the opaque bars of a transmission grating are made as reflecting or mirror bars, the a moving image on such a grating is split into transmitted and reflected parts. Then, the intensity of reflected light varies periodically as that of transmitted light in the normal transmission grating velocimeter and is used for SFV measurements. This type of device may be called a mirror grating. Figure 5.21 shows an example of the SFV optical system using a mirror grating. In this example, the mirror grating is formed by deposition of a cube beam splitter on the exit plane, and the moving image is focused on this plane. The light reflected by periodic mirror bars received by photodetector PD_2 whereas the transmitted light is received by PD_1. The two periodic signals from PD_1 and PD_2 contain the same central frequency, but the phase difference π. Thus, the differential output of the two signals has no pedestal component. This system realizes SFV measurements with pedestal removal by using a single optical device; a beam splitter with a mirror grating. Then, the system is expected to be a little more compact and stable, but the mirror-grating type of beam splitter or even just a simple mirror grating is not a general optical component. This type of grating may have to be designed and fabricated as occasion demands.

5.3 Lenticular Grating

A lenticular grating consists of an array of small cylindrical lenses, as shown in Fig. 5.22. This optical device is commercially available as a lenticular lens, a

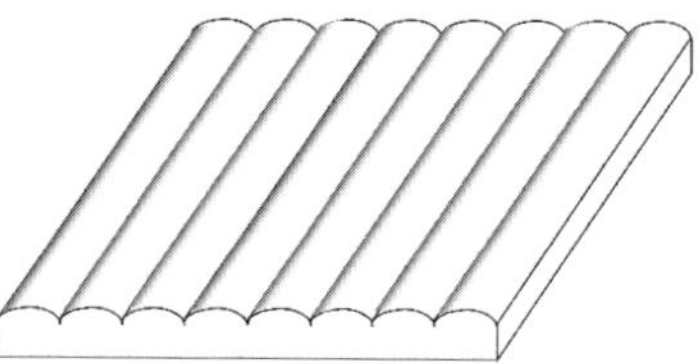

Fig. 5.22. Lenticular grating

kind of Fresnel lens, and its typical ranges of grating pitch and the focal length are about $200 \sim 500\,\mu$m and 300 μm ~ 2 mm, respectively. The lenticular lens was used to process old color photographic films of Kodak and is now mostly used in stereophotography or stereograms. The light incident on the flat face of the lenticular lens is refracted at the exit interface due to the spherical surface of cylindrical lenses. Thus, the lenticular lens is used for a spatial filtering device based on a principle similar to that of the prism grating.

5.3.1 Lenticular Grating Velocimeter

Figure 5.23 [42] shows schematically the basic construction of a SFV system using a lenticular grating. The light scattered by a moving object forms an image of the object on the flat face of the lenticular grating (LG) by means of lens L_1. A mask M is attached to the flat face of the grating to define the probe cross-sectional area. The light passing through the grating is deflected into a range of angles from $-\theta_m$ to θ_m, as shown in Fig. 5.24 [42]. The deflection angle θ varies continuously according to the image movement along the x axis on the flat face of the grating. Since cylindrical lenses are arrayed on the x axis with an equal pitch, a change in the deflection angle is periodically repeated as the image moves. Two collecting lenses L_{21} and L_{22} evenly share the range of angles for receiving the light deflected from the grating and propagate the light alternately into photodetectors PD_1 and PD_2. This operation generates two periodic detector outputs of the same central frequency but with a phase difference of π. Thus, differential detection of the two outputs gives an SFV signal with the pedestal component removed, which can be used for velocity measurements.

In the same way as the prism grating, the lenticular grating has no significant loss of light for detection because of its almost transparent material. In Fig. 5.24, the negative and positive ranges of the deflection, from $-\theta_m$ to the z axis and from the z axis to θ_m are exactly symmetrical in angle only for the cylindrical lens CL_0 on the z axis. For other cylindrical lenses $CL_{m|m\neq0}$ off the z axis, the two ranges become asymmetrical and are not equal in angle. Then, the time periods for which the photodetectors PD_1 and PD_2 receive collected light become different, whereas the time intervals of the two periodic signals are still the same. This effect may cause undesirable higher frequency components in the differential output. Since the effect is due to the location

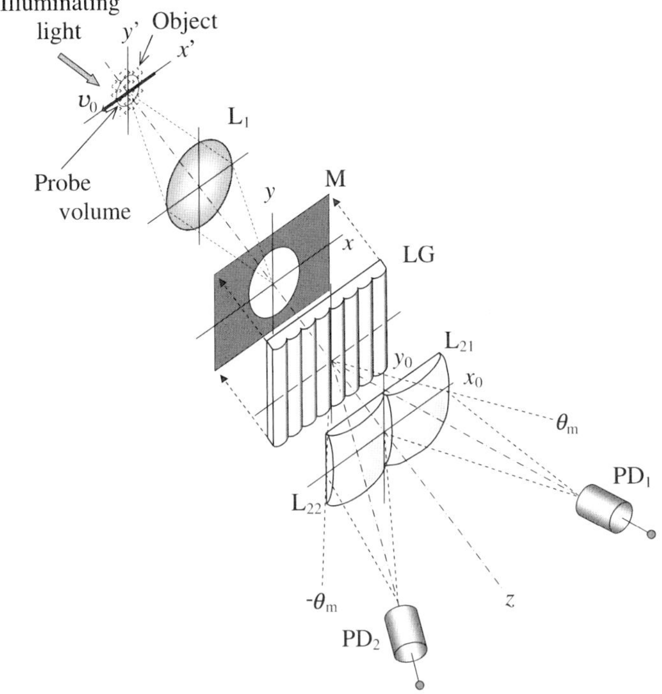

Fig. 5.23. Basic construction of a lenticular grating velocimeter [42]

of cylindrical lenses out of the z axis, the width of each cylindrical lens or the grating pitch p must be small enough to be neglected in comparison with the two distances between the imaging lens L_1 and the grating LG and the grating LG and the plane of the collecting lenses L_{21} and L_{22} [42]. The number of grating lines, n, is given by that of the cylindrical lenses included in the mask M. An unnecessarily large number n may also cause the above effect.

The visibility of the output signals may depend on two conditions in the lenticular grating velocimeter. One is the ratio of the image size $2b$ of an individual object such as a particle to the interval p of cylindrical lenses or the grating pitch. As described in Sect. 2.6.2, the ratio $2b/p$ is a primary factor deciding signal visibility and is common to all types of SFV systems. High visibility can be obtained for a smaller value of $2b/p$, such as ≤ 0.5. Another parameter is the ratio of the diverging angle θ_a of the light flux deflected from the grating to the total deflection angle $2\theta_m$. This is because periodic signals are generated by the fact that the light flux spread by θ_a scans two collecting lenses L_{21} and L_{22} periodically within $2\theta_m$. Visibility as a function of the ratio $\theta_a/2\theta_m$ is derived in the same way as that for the ratio $2b/p$ in Sect. 2.6.2 and is given for the one-dimensional case by [42]

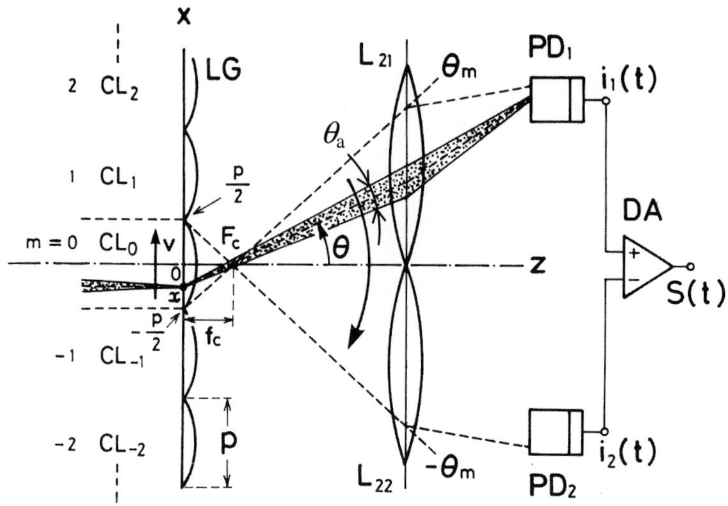

Fig. 5.24. Deflection of light by the lenticular grating [42]

$$V = \left| \frac{\sin \pi \left(\dfrac{\theta_a}{2\theta_m} \right)}{\pi \left(\dfrac{\theta_a}{2\theta_m} \right)} \right| . \tag{5.16}$$

This equation is a well-known sinc function and it can be estimated that high visibility is obtained for $\theta_a/2\theta_m \leq 0.5$. This condition should be accounted for in designing the optical system for lenticular grating velocimeter.

5.3.2 Directional Discrimination

In comparison with a prism grating, a specific feature of the lenticular grating is flexibility in the number of detection channels. Although the prism grating is followed by two or three detection channels according to the number of stages (two or three, and possibly a few more), the detection channels for the lenticular grating can be chosen by varying the number of collecting lens/detector sets introduced in the deflection range of angles $2\theta_m$. Figure 5.25 illustrates an example of four detection channels in a lenticular grating velocimeter by using four identical sets of collecting lenses and photodetectors. In this example, four outputs S_1, S_2, S_3, and S_4 from the detectors contain the same central frequency proportional to the object's velocity, with a phase difference of $\pi/2$ between neighboring detectors. Note that these outputs also include a pedestal component. However, a simple operation can be applied to the four outputs to generate new signals S_a and S_b, as

$$S_a = (S_3 + S_4) - (S_1 + S_2) , \tag{5.17}$$

$$S_b = (S_4 + S_1) - (S_2 + S_3) . \tag{5.18}$$

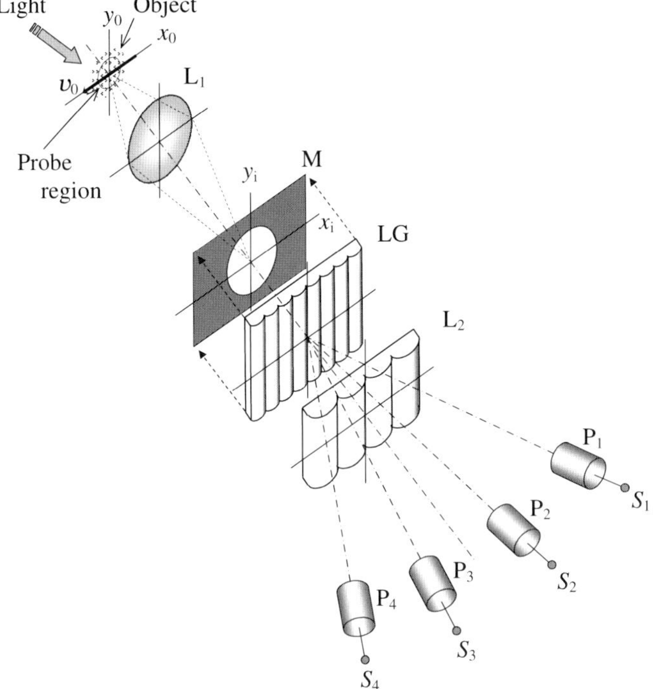

Fig. 5.25. Example of four detection channels in a lenticular grating velocimeter

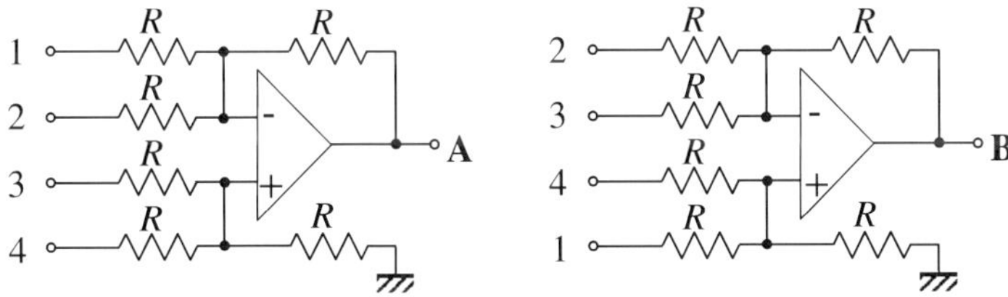

Fig. 5.26. Basic operating circuit for directional discrimination using four detection channels

This processing is done by a basic operating circuit such as Fig. 5.26, in which the numbers **1** to **4** denote four detector outputs S_1 to S_4, and **A** and **B** indicate resultant signals S_a and S_b, respectively. The two signals S_a and S_b still keep the original central frequency with a phase difference of $\pi/2$ and, this time, their pedestal component has been removed by subtraction in the above operations. Since the sign of the phase difference is decided by the direction of a moving image on the lenticular grating, directional discrimination can be done by comparing the two signals S_a and S_b, for instance, in the signal processing of Fig. 5.13, or by using a determination of the instantaneous frequency of the analytic signal (see Sect. 5.8).

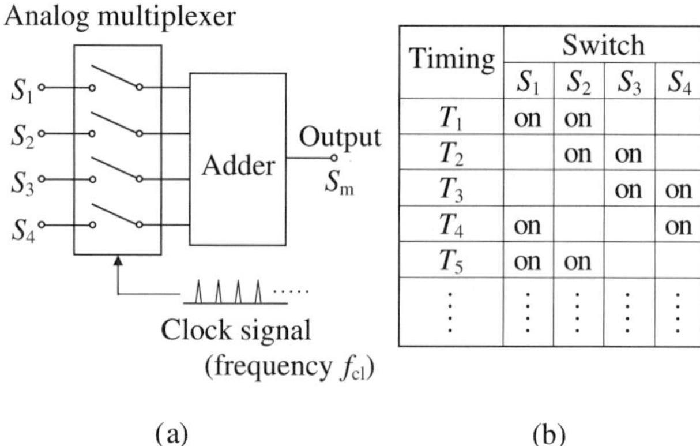

Timing	Switch			
	S_1	S_2	S_3	S_4
T_1	on	on		
T_2		on	on	
T_3			on	on
T_4	on			on
T_5	on	on		
⋮	⋮	⋮	⋮	⋮

(a)   (b)

Fig. 5.27. Concept of frequency shifting for directional discrimination in a lenticular grating velocimeter

The above method is based on the phase shifting technique. Although the lenticular grating itself cannot be moved, equivalent effect is produced by an appropriate signal processing and, thus, the frequency-shifting technique can also be realized to determine the direction of the object's movement in a lenticular grating velocimeter of Fig. 5.25. An example of such processing is conceptually described in Fig. 5.27. Four detector outputs are fed into an analog multiplexer, which switches two adjacent outputs on and the rest off in the first period of time and, then, shifts the connection one by one in each period in the successive timing states. Any two of the four outputs selected by the multiplexer are added to produce the signal S_m. Figure 5.27b demonstrates that the signal S_m equals $S_1 + S_2$ at time T_1, $S_2 + S_3$ at T_2, and so on. This is equivalent to the transmittance of a moving grating as shown in Fig. 5.28. In this way, the effect of a moving grating can be electronically realized for a lenticular grating. In the example of Fig. 5.27, one cycle of the signal S_m consists of four timing periods. Then, if the multiplexer shifts the connection at a clock frequency f_cl, the shift frequency f_g of the equivalent moving grating is given by $f_\mathrm{cl}/4$. Therefore, the direction of the object's movement can be determined by assessing whether the central frequency of the signal S_m is higher or lower than the shift frequency f_g, based on the principle of the frequency-shifting technique described in Sect. 5.1.3.

5.3.3 Two-Dimensional Measurements

A further development of the lenticular grating velocimeter is to measure velocity in two dimensions. Figure 5.29 shows the basic optical system for measuring two-dimensional velocity components using two lenticular gratings which are placed orthogonally in the image plane. The first lenticular grating LG_1 deflects the transmitted light into a range in the x direction with an image movement along the x axis whereas the second grating LG_2 deflects

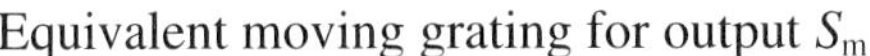

Equivalent moving grating for output S_m

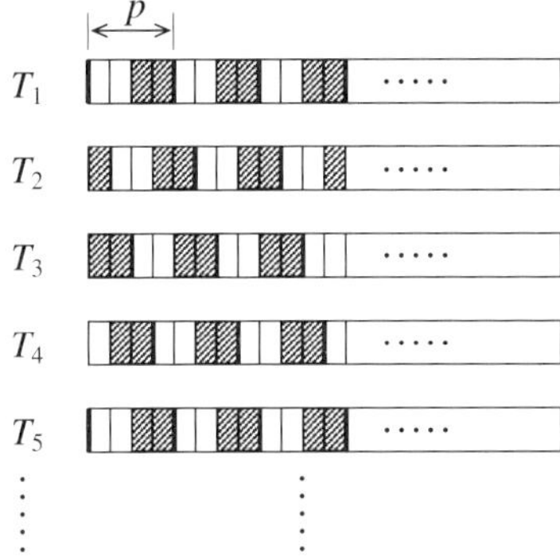

Fig. 5.28. Equivalent effect to the transmittance of a moving grating

the light in the y direction with the movement along the y axis. The light deflected in both x and y is collected by a set of four collecting lenses L_2 and received by photodetectors P_1 to P_4. In the same way as directional discrimination using four detection channels, the detector outputs are again fed into the operating circuit of Fig. 5.26, from which two periodic signals S_a and S_b given by (5.17) and (5.18) are obtained, respectively. In the detector configuration of Fig. 5.29, the signal S_a contains the central frequency f_x that is proportional to the x component of the object's velocity, whereas S_b has f_y for the y component. The pedestal component is eliminated in S_a and S_b by subtraction in (5.17) and (5.18). Therefore, measurements of two central frequencies f_x and f_y provide two-dimensional velocity components. Although ratio of the two components gives an argument of the object's velocity, the directional ambiguity of 180° still remains in the system of Fig. 5.29.

To perform directional discrimination or two-dimensional measurements, four detectors are used in the above examples. If necessary, the number of detectors can be increased in principle in the lenticular velocimeter. The use of multiple detectors may cause additional problems in the system, such as the need of balancing the sensitivity of all the detectors and making the system compact. However, these may not be serious problems if, for example, a quadrant-type photodiode or multianode photomultiplier is employed in the construction of a practical system.

5.4 Optical Fiber Array

The end faces of optical fibers which are linearly arrayed at equal intervals can be used as a spatial filtering device for velocity measurements. Since an optical fiber is one of the wave guiding devices, it flexibly connects an optical imaging system with a photodetector. In addition to this flexibility, the

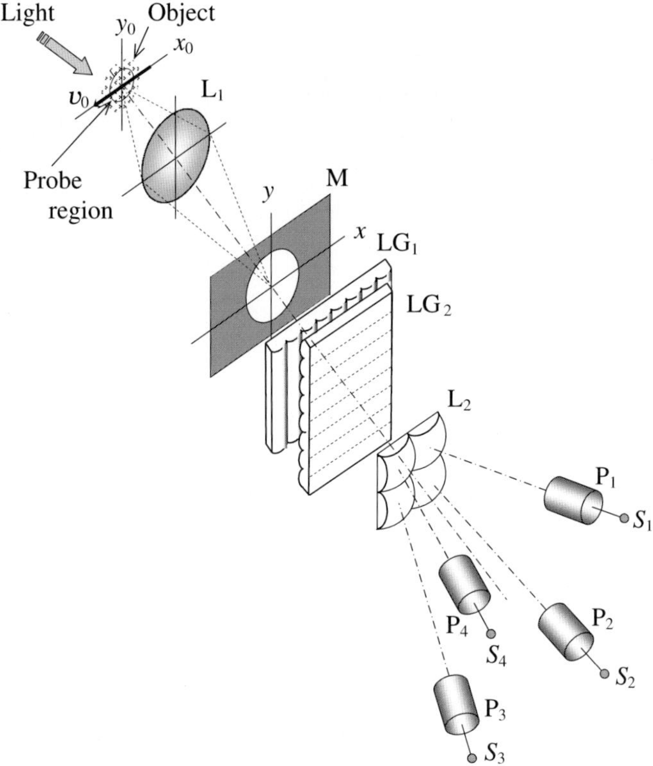

Fig. 5.29. Configuration of a lenticular grating velocimeter for two-dimensional measurements

optical fiber has practical advantages such as electromagnetic noise tolerance and high dielectric strength. In the SFV system using an optical fiber array, the intensity of light must be propagated through the optical fiber, but the phase information of the light field is not required, and, thus, various types of commercial optical fibers are available at inexpensive prices for the spatial filtering device.

5.4.1 Optical Fiber Array SFV

Figure 5.30 shows schematically the basic construction of the SFV optical system using an optical fiber array [32]. The image of a moving object is formed by lens L at the input faces of the fibers and is guided through the fibers to two photodiodes PD_1 and PD_2. In a spatial filtering device of this type, the transmittance as a spatial filter is decided by the pattern of the fiber array being constructed, assuming that all fiber cores used have the same and uniform transmittance. Although a variety of array patterns can be designed, the basic pattern is the construction illustrated in Fig. 5.31 [32]. The input

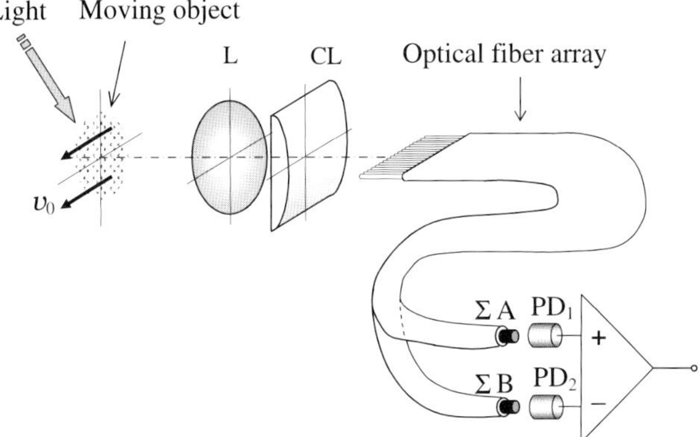

Fig. 5.30. Basic construction of a SFV using an optical fiber array

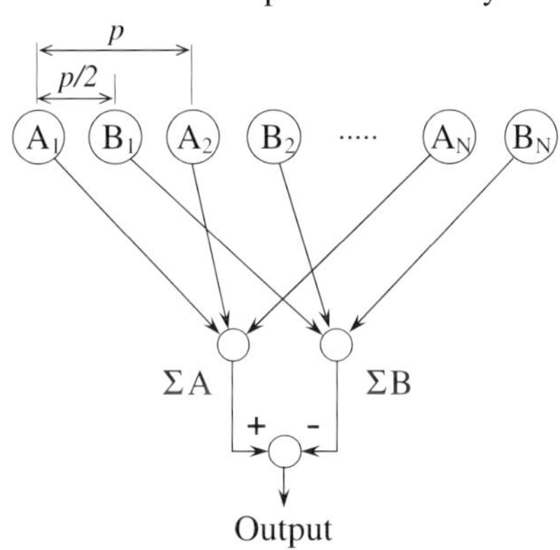

Fig. 5.31. Basic connection pattern of fibers and photodetectors [32]

faces of fibers having an equal diameter are linearly arrayed at intervals of $p/2$. On their output sides, every other fiber is bundled, and their end faces look toward photodetector surfaces. By this construction, the light from every other fiber is collected and detected to produce two periodic detector outputs having the same central frequency with a phase difference of π. The two outputs are fed into a differential amplifier, from which the periodic signal without a pedestal component is obtained. In this way, the fiber array of Fig. 5.31 acts as a differential-type spatial filter, playing the roles of two transmission gratings, beam splitter, and waveguides. The grating pitch is given by p in the above example.

The linearly arrayed fibers sense the image velocity along the line of the array. If an object moves in a direction tilted from this line, the moving image is not detected by all fibers and output signals may deteriorate. It is quite easily imagined when one considers generally very small diameters of fiber cores. To solve this problem, a cylindrical lens (CL) is inserted behind lens L as shown in Fig. 5.30. By this imaging system, the image intensity distribution is averaged in a direction perpendicular to the fiber array, and any sideways motion of an object is effectively detected by all fibers.

The filtering characteristics of an optical-fiber-array spatial filter are determined mainly by the fiber interval (grating pitch), core diameter, and the number of fibers used in the array. They have been theoretically elucidated [32] and may equivalently be estimated by those of the transmission grating which are described in Chap. 2. Although there is difficulty in constructing an array precisely in a micron order, it is a practical merit that the optical system and the detector/signal-processing unit can be separated.

5.4.2 Directional Discrimination and Two-Dimensional Measurements

Analogous to other spatial filtering devices, two developments in the optical fiber array can easily be made to achieve directional discrimination and two-dimensional measurements in a SFV system. Figure 5.32 [35] illustrates the configuration of an optical fiber array for velocity measurements having directional discrimination. In this regime, the light from every fourth fiber is collected and detected by four photodetectors, which produce four outputs having the same central frequency with a phase difference of $\pi/2$, together with the pedestal component. This situation is equivalent to that obtained in the lenticular grating velocimeter with four detection channels, shown in

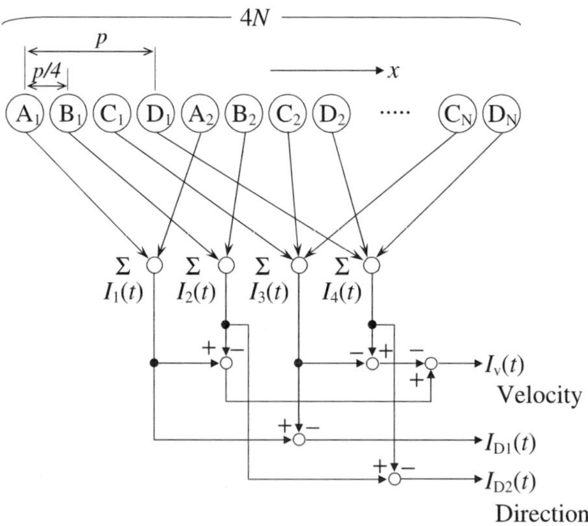

Fig. 5.32. Connection of fibers and photodetectors for the directional discrimination [35]

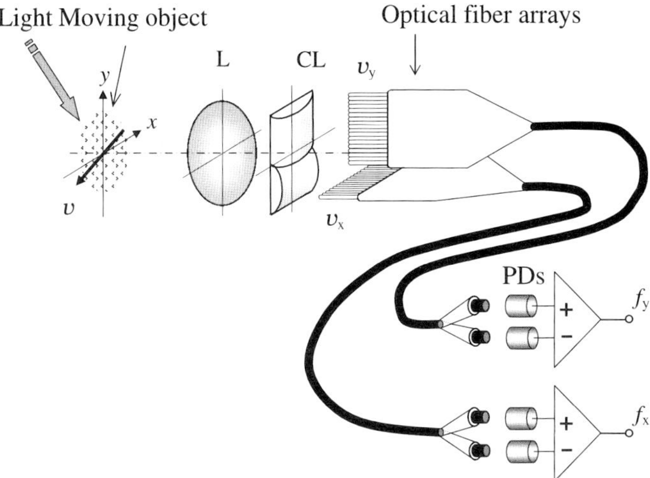

Fig. 5.33. Configuration of two optical fiber arrays for two-dimensional measurements

Fig. 5.25. Then, simple signal processing such as that given by (5.17) and (5.18) or shown in Fig. 5.26 is used to sense the moving direction. The grating pitch p in this type of fiber array is four times as long as the fiber interval. When a smaller value of p is difficult to obtain because of the lower limit of fiber diameters, the image of the moving object being measured may be magnified according to requirements. By treating the phase-orthogonal output pair as an analytic signal, the instantaneous phase, frequency, and, thus velocity can also be obtained with the sign giving the moving direction.

Two-dimensional velocity components can be measured by a pair of optical fiber arrays which are orthogonally placed in the image plane [33]. An example of such a configuration is schematically shown in Fig. 5.33. Two cylindrical lenses, which are inserted into the diffraction region of the imaging system, share the imaging light flux and form two moving images having different intensity distributions. One is averaged in the y direction to detect the x component of the velocity and another is done in the x direction for the y component. Each optical fiber array transmits light to two photodetectors that are differentially connected and generates periodic signals whose central frequency is proportional to the x or y velocity components. The range of averaging in the image plane may be adjusted by the focal length and position of the cylindrical lens to cover the extent of object distribution or the probe cross-sectional area.

5.5 Liquid Crystal Cell Array

Liquid crystals (LCs) have long cigar-shaped molecules, which move like liquids or solidly interact to form a structure like crystals. These physical properties between liquids and solids, which are controlled by applying a voltage, show interesting optical phenomena such as changes in polarization, phase, and transmittance and scattering. As a typical example, LCs are used as optical devices that switch their transparent/opaque states by a controlling voltage. By using this optoelectric property, a new type of spatial filtering device can be developed. Since there are potentially various schemes for controlling voltage, this type of spatial filter possesses much flexibility in its spatial and temporal filtering characteristics. Although LC spatial filters require the application of voltage, the power consumption is generally low enough for practical uses. The disadvantages of the LC cell array are its low frequency response, low contrast and optical properties that are sensitive to a temperature change.

5.5.1 Liquid Crystal Spatial Filter

When an ac electric field is applied to a transparent LC cell, it becomes opaque due to its property of dynamic light scattering. This type of LC is able to alternate its transparent and opaque states without polarizers. Figure 5.34 [21] shows the construction of a LC spatiotemporal spatial filter developed by Itakura et al., which realizes a time-varying spatial transmittance distribution. A thin nematic LC layer is enclosed between an optically flat transparent electrode plate and an array of small strip electrode cells. Each strip cell can be switched electrically and independently on or off by a driving circuit. Figure 5.35 [21] illustrates the principle of the time-varying transmittance distribution realized by the proposed LC spatial filter. In this example, four neighboring cells form one group which corresponds to a transparent or opaque bar of the general type of parallel-slit reticle. The ac electric field is applied (switched on) to every other group to construct equivalently a periodic arrangement of opaque bars. Other groups to which the electric field is not applied (or switched off) equivalently act as periodic transparent bars. By shifting the cells forming the opaque and transparent groups electrically and sequentially

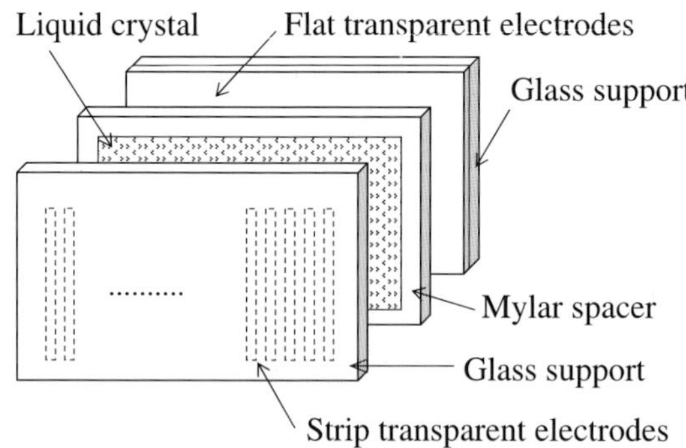

Fig. 5.34. Construction of the liquid crystal spatial filter (re-drawn from [21])

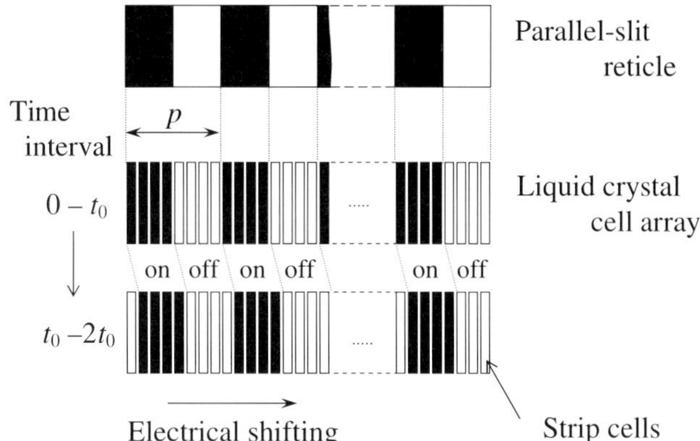

Fig. 5.35. Principle of time-varying transmittance distribution realized by a LC spatial filter (re-drawn from [21])

one by one with a time interval t_0, the transparency/opaqueness pattern or equivalent transmission grating pattern moves intermittently. Therefore, the spatial filter is realized with time-varying spatial transmittance distribution, which provides the frequency-shifting technique to determine the velocity direction without any mechanical movement. In Fig. 5.35, the grating pitch p is given by the width of eight strip cells. The number of cells within one group may be changed by altering a driving scheme in the circuit according to requirements. Therefore, pitch p is electrically variable. The shifting frequency or moving velocity of the time-varying transmittance pattern is controllable by the shifting time interval t_0 in the circuit.

The shifting behavior of the LC spatial filter is not continuous but intermittent, as mentioned above. This intermittence may affect the properties of output signals. Theoretical investigations indicate that the effect is negligible for a larger number of grating lines and the number of cells within one group larger than four [21]. This condition makes the effect of discrete group movement rather unnoticeable. Another problem that should be addressed is the finite response time of LC cells, which corresponds to the finite rise time and fall time for the dynamic light scattering effect of the LC device caused by switching the voltage supply on and off. This phenomenon indicates that the contrast of a time-varying opaqueness/transmittance pattern is degraded by a higher repetition rate of the voltage supply. Thus, the repetition period must be set larger than the expected rise or fall time of the device being used. On the other hand, the above phenomenon fortunately works to smooth out the intermittent behavior of the LC spatial filter, and the defect is relieved.

5.5.2 Piled Construction for Velocity-Vector Measurements

When two LC spatial filters are piled orthogonally in the lines of their strip cells, a two-dimensional velocity vector can be measured since each LC spatial filter discriminates the 180° velocity direction in its dimension. The glass support and Mylar^{TM} spacer are usually made thin enough to neglect thickness of the piled LC cell arrays (1 mm for Itakura's device [21]) in comparison with the focusing depth of the moving image. Then, the light passing through the first LC spatial filter enters the second filter straight within the imaging plane. With this construction, two-dimensional measurements are realized in a single optical imaging system with a single photodetector. The detector output is processed by the spectrum analyzer since two periodic components of x and y are mixed in it. To determine each central frequency of the two components independently, the two shifting frequencies for the LC cell arrays should be set to be different with sufficient separation in frequency.

Since the piled construction of LC spatial filters provides two-dimensional velocity-vector measurements, it has potential for practical uses. Although the frequency response, contrast, and spatial resolution limit its application, recent developments in LC materials and devices are expected to improve the prospects for LC spatial filters.

5.6 Integrated Solar Cell Array

The spatial filtering devices described so far in the previous five sections are used as optical elements that are placed in the image plane between the imaging lens and the photodetectors. This and the next sections treat a detector type of spatial filtering device, which operate both as a spatial filter and as a photodetector. The detector type usually contributes to simplification of the optical system since the light is detected directly in the image plane and the device requires no additional collecting optical system. However, photodetectors used for this purpose are generally sensors with insufficient sensitivity and data rate. Thus, the detector-type spatial filter is effective for low-velocity moving objects from which sufficient intensity of scattered light is obtained. Recent developments in the array type of detectors may improve the outlook.

5.6.1 One-Dimensional Array

An integrated array of silicon solar cells having equal pitches is a typical and primitive example of a detector-type spatial filtering device. Figure 5.36 [52,107] shows schematically the construction and principle of a device built on a ceramic substrate. This integrated array is made by etching a surface pattern of p-type silicon on an n-type silicon base. The etched pattern forms an array of minute p-type silicon cells as a parallel-slit reticle. Each minute cell generates an electric current which is proportional to the light intensity of the projected moving image. The current from each cell is collected by an aluminum electrode connected electrically to a side of the cells and is led via a load resistor to an amplifier. By this construction, the detector surface acts as a photodetector and also as a spatial filter. The example of Fig. 5.36a contains

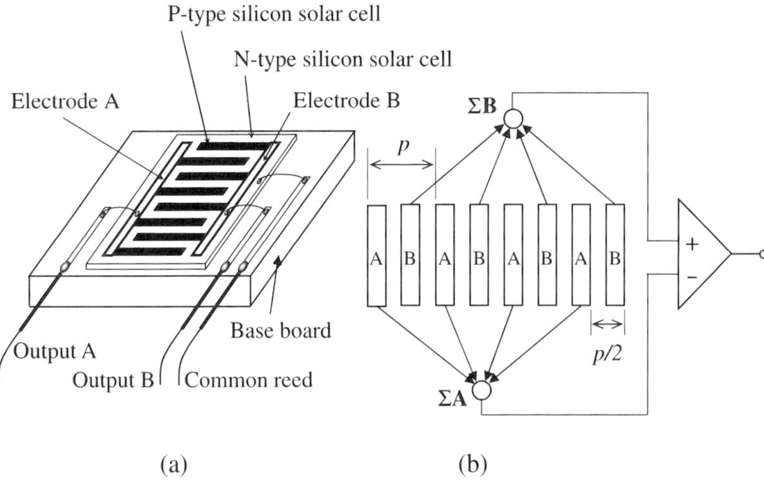

Fig. 5.36. A detector-type spatial filtering device using an integrated solar cell array [52, 107]

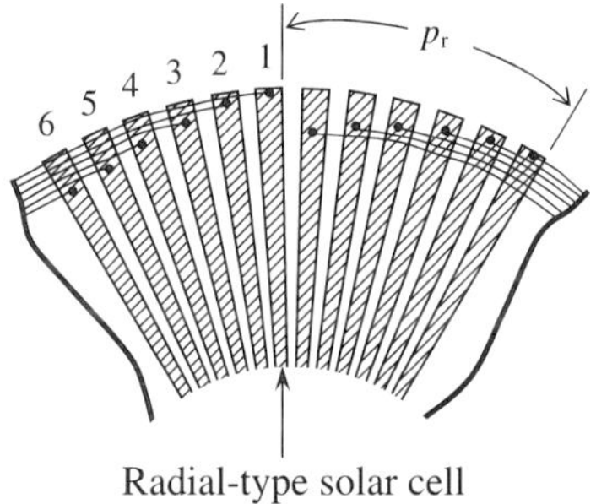

Fig. 5.37. Radial-type solar cell array designed for measuring rotating velocity [19]

two channels A and B of solar cells, which are alternately arrayed to realize differential detection, as shown in Fig. 5.36b. Thus, output signals from the differential amplifier have no pedestal component and can be analyzed directly in the signal processing system to determine the object's velocity.

To fit the device to different types of object motion, some other etched patterns have been developed. Figure 5.37 [19] shows the construction of a radial-type solar cell array having equal angular pitches, which has been designed for measuring rotating velocity. In this example, every sixth cell is electrically connected to generate one output, and totally six channels of output signals are obtained with a phase difference of $\pi/3$, which may be effectively used in signal processing to eliminate the pedestal component and to discriminate the direction of rotation. The central frequency f_{0r} of each of six signals is given theoretically by

$$f_{0r} = \frac{\omega_0}{p_r} , \tag{5.19}$$

where ω_0 and p_r denote the angular velocity (rad/s) of a rotating object and the angular pitch (rad) of the radial cell array. Note that the angular pitch p_r of the example in Fig. 5.37 is given by the angle covered by six adjacent radial cells. When the total number of cells in the radial cell array and the number of detection channels are denoted by N and n_{ch}, respectively, the effective number of grating lines, n_r, for each output signal is expressed as

$$n_r = \frac{N}{n_{ch}} . \tag{5.20}$$

Then, the angular pitch p_r is given by

$$p_r = \frac{2\pi}{n_r} = \frac{2\pi n_{ch}}{N} , \tag{5.21}$$

and the central frequency f_{0r} is rewritten as

$$f_{0r} = \frac{n_r}{2\pi}\omega_0 = \frac{N}{2\pi n_{ch}}\omega_0 . \tag{5.22}$$

Equation (5.19) or (5.22) indicates that the central frequency f_{0r} is independent of the imaging magnification, and this is the unique feature of the radial cell array.

A problem which is peculiar to the radial type of cell array is an error in alignment between the two center axes of the radial cell array and the projected image of the rotating object. The error reduces the intensity of the periodic signal component having the central frequency f_{0r} and, in exchange for that, undesirable sidelobe components appear in the frequency spectrum. Using outer areas of the radial cell array may moderate the above defect.

Another example of integrated solar cell patterns is the herringbone array [52] shown in Fig. 5.38, which was originally proposed by Ator [12], as shown in Fig. 5.17, to sense the drift of a moving direction. Two solar cell arrays (α and β) tilted at angles, $\pm\phi$ are symmetrically arranged with respect to the centerline. Each array produces two outputs from the electric connection of every other cell and realizes differential detection for removing the pedestal.

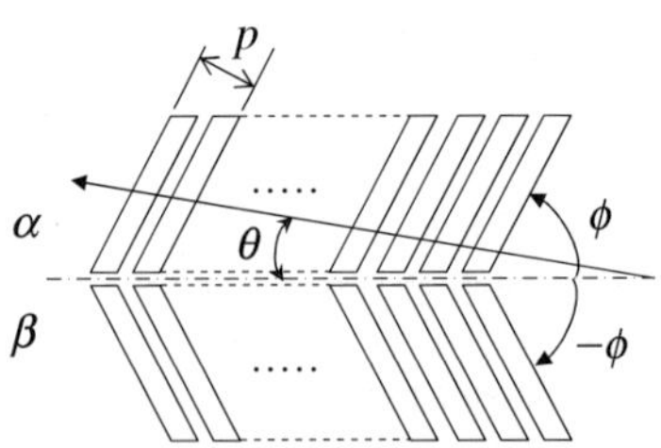

Fig. 5.38. Herringbone pattern of the integrated solar cell array

When the central frequencies of two resultant periodic signals from arrays α and β are measured as f_α and f_β, the object's velocity v and the directional angle θ from the centerline are determined by

$$v = \frac{p}{M}\sqrt{f_\alpha{}^2 + f_\beta{}^2}\,, \qquad (5.23)$$

$$\theta = \tan^{-1}\left(\frac{f_\alpha - f_\beta}{f_\alpha + f_\beta}\right)\,. \qquad (5.24)$$

These equations are derived for the representative case that the angle ϕ is assumed to be $\pi/4$ for mathematical simplicity. As seen from (5.24), $f_\alpha = f_\beta$ means that $\theta = 0$, which corresponds to a moving direction along the centerline. Thus, the herringbone array can be used for sensing the drift angle of a moving object. If the angle $|\theta|$ approaches $\pi/4$, one of the two arrays generates almost no periodic signal since the moving image does not cross solar cells of the array but travels in parallel with them. For $|\theta| > \pi/4$, the array produces erroneous periodic signals, which do not give true values of v and θ in (5.23) and (5.24). Thus, there is a range of drift angles that can be measured with the device of Fig. 5.38.

5.6.2 Two-Dimensional Array

There is also an example different from the previous three cell arrays. Figure 5.39 [108] shows the basic construction of an integrated-cell-array spatial filter with weighting variability. Although the device developed has 256 pixels of photodiode cells in a 16×16 arrangement on a silicon substrate of a

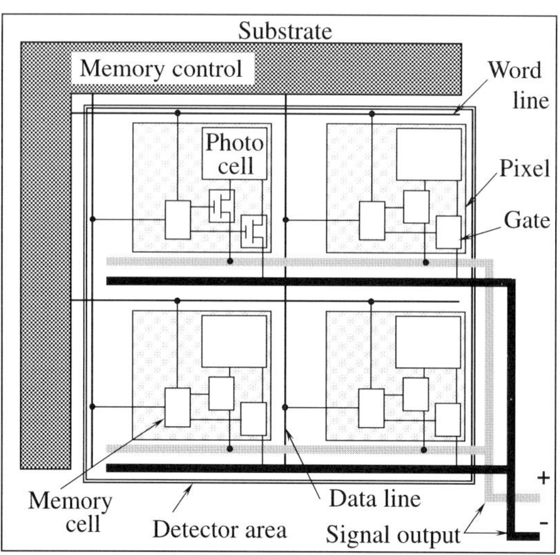

Fig. 5.39. A two-dimensional array of integrated solar cells for a spatial filter [108]

one-chip solid-state element, Fig. 5.39 illustrates part of four pixels in 2×2 for schematic simplicity. Each pixel consists of one photodiode cell, two FET gates, and a 1-bit memory cell. An output from the photodiode cell is connected via the two gates to two signal lines of positive $(+)$ and negative $(-)$ outputs. According to the weighting data stored in the memory cell, one of the two gates opens to transfer the photodiode output to the signal line and the other closes. Writing the weighting data into the memory is done pixel by pixel in terms of word and data lines which are connected with an external circuit for setting the weighting function. The two signal lines $(+)$ and $(-)$ are connected to a differential amplifier. Thus, pixels connected to the positive line are weighted by $+1$, and those to the negative by -1. The weighting pattern of total pixels can be changed electrically by controlling the weighting data. If this device is used as a spatial filter, the effective pitch and direction of grating lines become variable. Then, the device possesses adaptivity to the image condition (moving direction and light intensity distribution) of an object being measured. The device presented here is a beginning-stage example of this type, and there are potentially various avenues for its development and application in the future.

5.7 Line Sensor

In general, integrated solar-cell arrays are not commercially available and must be designed and fabricated in compliance with requirements. This section introduces the use of commercially available one-dimensional array devices for detector-type spatial filters. In addition to this, two-dimensional or areal sensors open new prospects for the detector-type spatial filter with advanced electronic circuits and/or computer image processing. These approaches will be presented in Sect. 5.8, separately from this section.

5.7.1 Linear Photodiode Array

An example of such devices is a linear photodiode array [109, 110] placed in the image plane. For detecting translational speckles, the photodiode array

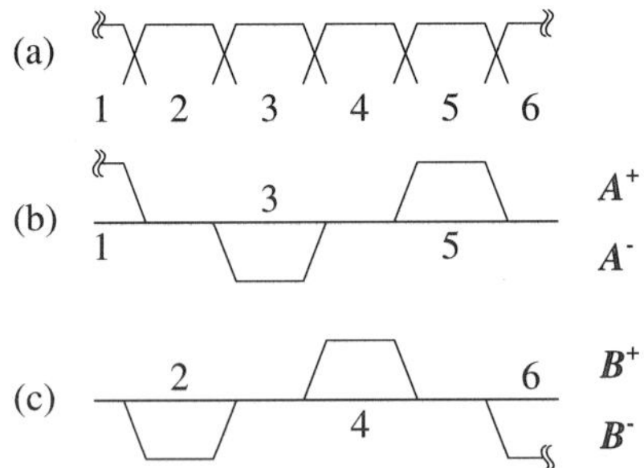

Fig. 5.40. Spatial response function of a linear photodiode array and the photo-current output (after [109])

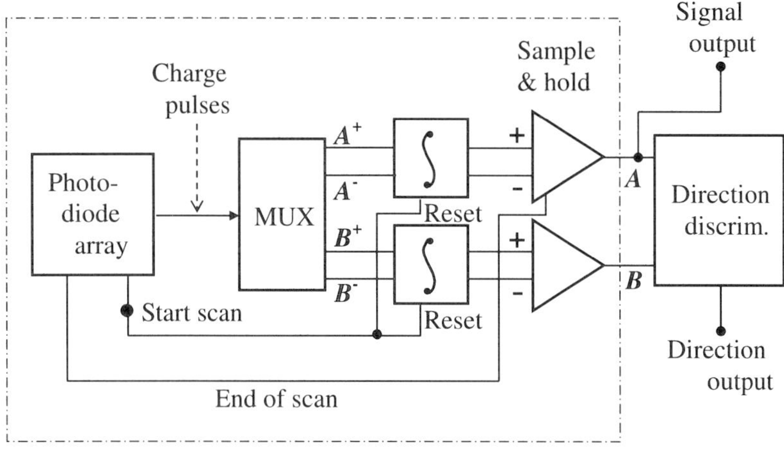

Fig. 5.41. Signal processing of a linear photodiode array used as a spatial filter (after [109])

may also be set in the diffraction plane [111]. When a particle image, which is assumed to be smaller than each diode cell, traverses a linear photodiode array having a spatial response function [109], as shown in Fig. 5.40a, a photocurrent output from each diode is read out sequentially by scanning all diode cells electronically. If the weighting function is electronically arranged to produce four periodic photocurrents, as shown in Figs. 5.40b and c, each current consists of outputs from every fourth diode, and the four photocurrents contain the same central frequency with a phase difference from each other. Figure 5.41 [109] shows a schematic diagram of an electronic signal-processing system for the device. The individual diodes are clocked into a demultiplexer which generates the above-mentioned four current outputs A^+, A^-, B^+, and B^-. Two pairs of A^+, A^- and B^+, B^- are differentially amplified to produce two signals A and B, respectively, whose pedestal components cancel out. At the end of scanning, signals A and B are used to discriminate the moving direction since the two signals are $\pm\pi/2$ out of phase. The object's velocity is determined by measuring the central frequency of signal A or B. In this way, a detector-type spatial filter with pedestal removal and directional discrimination is electronically realized by using a linear photodiode array. The main advantage is that the weighting function can be flexibly designed in the electronic system, so that the grating pitch is variable. A large number of diode cells in the array increases the scanning time and, thus, imposes a limitation on the velocity being measured. The detecting sensitivity of the diodes becomes a problem for measurements at a low light level. In this condition, multianode photomultiplier tubes can be considered.

The same type of device was studied [112] for laser speckle velocimetry with a special relation to the statistical properties of amplitude and phase of

SFV signals. In this study, the linear photodiode array was referred to as a differential comb photodetector array.

5.7.2 CCD Line Sensor

In the same way as the photodiode array, a charge coupled device (CCD) can also be used as a detector-type spatial filter [62, 113–115]. From recent developments in faxes, scanners, and other imaging systems, CCDs have become very popular as one- and two-dimensional (line and area, respectively) sensors, and a variety of models are provided in the market. The use of a CCD line sensor for a spatial filtering device is based on the same operation as the linear photodiode array. As shown in Fig. 5.42 [114], outputs from pixels of the CCD line sensor are divided into four channels in the sample-&-hold circuit by clock pulses that are synchronized with the reading clock for the pixels. The four channels of signals S_1, S_2, S_3, and S_4 are produced by contributions from every fourth pixel and processed to make two differential outputs S_a and S_b based on (5.17) and (5.18), apart from the order of subtraction. The resultant S_a and S_b can be used for discriminating the moving direction as well as determining the velocity. Unless the sense in the moving direction is necessary, the system can be constructed for two channels of the signal, which correspond to outputs from every other pixel and contain the phase difference of π from each other. Then, differential detection can be performed to remove the pedestal component.

The problem of readout time should be discussed in the use of CCDs. Consider a simple case by assuming a CCD line sensor with 1024 pixels and a maximum clock frequency of 1 MHz. To read out all pixels, it takes nor-

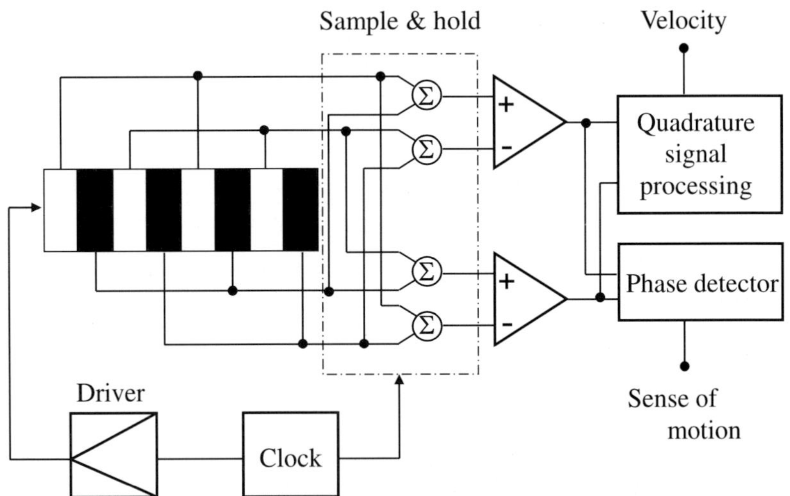

Fig. 5.42. Signal processing of a CCD line sensor used as a spatial filter [114]

mally $1024\,\mu s$, and, thus, the data rate in this case is only about $1\,KHz$ at most. The readout time limits the maximum detectable velocity. In driving the CCD sensor, there is a useful operation, in which summations are carried out by dynamic charge accumulation within the transport section of the CCD line. This operation is known as "pixel binning" and can be achieved by so-phisticated clocking [62,114]. By using pixel binning, a maximum data rate of 500 kHz can be achieved in the above case and the problem may be improved.

A two-dimensional or area-type CCD sensor is used in the same way as an one-dimensional or line-type sensor when pixel binning is employed in a direction perpendicular to the moving direction. Assume that the object image is moving along the row of a CCD area sensor. Charges from the pixels of each column are integrated and read out as the output of a single equivalent pixel within the row. Thus the column acts as a single long pixel. This operation is advantageous for moving objects having a certain width because, due to the two dimensions, more light can be detected than that with the one-dimensional sensor. Some recent models of the CCD area sensor are equipped with a binning operation of this type.

5.8 Area Sensor and Video Camera

Sections 5.6 and 5.7 describe mainly the one-dimensional detector type of spatial filtering device. This section presents a two-dimensional detector type, in other words, an imaging type of spatial filtering device, which acts both as a spatial filter supported by electronics/computer and as an image detecting device. For this type of device, solid-state imaging devices as well as conventional imaging cameras may be available. The use of this type simplifies the measuring optical system since no specific optical component is required and the spatial filtering operation is performed fully in electronic circuits or by computer software. This implies that the parameters of the spatial filter, such as the pitch and number of grating lines, are variable. Another useful advantage of using the imaging type is that one can observe the images of moving objects during measurements. The temporal resolution of velocity measurements with the imaging-type spatial filter is determined by the frame rate and is generally insufficient for measurements of high-velocity objects.

5.8.1 Image Sensor with Electronic Circuits

An image sensor placed in the image plane receives the two-dimensional image intensity distribution of moving objects and converts it into sequential electric signals as a series of outputs from all pixels. If the signals are multiplied by an appropriate weighting function, which corresponds to a certain required spatial filter, the spatial filtering effect can be realized electronically [116–119]. In principle, this approach is the same as that used in one-dimensional sensors described in the previous section. Figure 5.43 [119] shows a schematic diagram

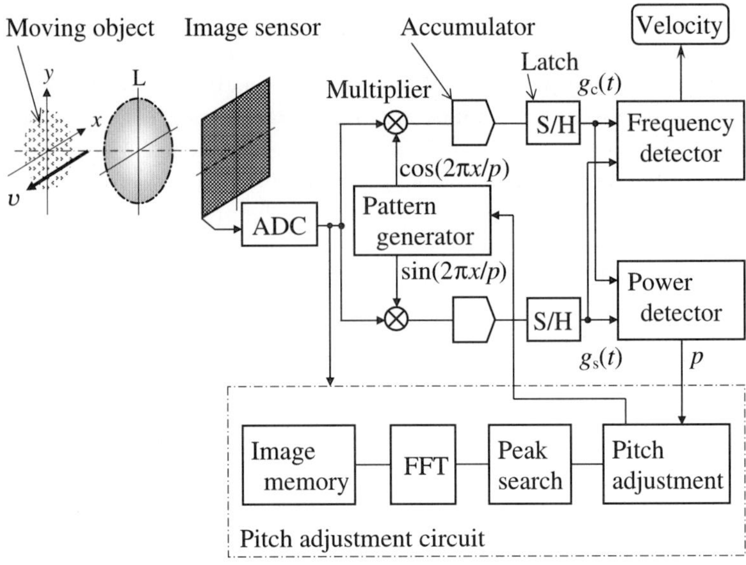

Fig. 5.43. Electronic spatial filter using an image sensor (re-drawn from [119], Copyright 1996, with permission from Elsevier)

of an electronic spatial filter using an image sensor with electronic circuits. The superposed weighting function, which is produced by a pattern generator, gives a two-dimensional transmittance pattern of the spatial filter having the same pixel numbers as those of the sensor image. Assume that the image intensity distribution of moving objects detected by the image sensor is denoted by $f(v_x t - x, v_y t - y)$, where v_x and v_y are velocity components of the image in the x and y directions, respectively. In the example of Fig. 5.43, the digital sensor signal is divided into two identical channels, and they are multiplied by two sinusoidal transmittance patterns $\cos(2\pi x/p)$ and $\sin(2\pi x/p)$ for intelligent operations which are described below. Here, p denotes the grating pitch. An accumulator then integrates the resultant signal in every frame and yields output signal $g_c(t)$ or $g_s(t)$, which is a value corresponding to the summation of all pixels in each frame time and expressed as

$$g_c(t) = \iint f(v_x t - x, v_y t - y) \cos\left(\frac{2\pi x}{p}\right) \, dx\, dy \,, \qquad (5.25)$$

$$g_s(t) = \iint f(v_x t - x, v_y t - y) \sin\left(\frac{2\pi x}{p}\right) \, dx\, dy \,. \qquad (5.26)$$

Due to these phase-orthogonal weighting functions, the two output signals $g_c(t)$ and $g_s(t)$ contain the phase difference of $\pi/2$, whereas their central frequencies are identical. Then, the object's velocity is determined by measuring the central frequency of $g_c(t)$ or $g_s(t)$ by usual signal-processing techniques which are described in Chap. 4. The temporal resolution is determined by the frame rate in this case. The moving direction can be discriminated, on the basis of the phase-shifting technique, due to the $\pi/2$ phase difference.

The advantages of the electronic spatial filter using an image sensor are as follows :

1. Various transmittance patterns for the spatial filter can be realized,
2. The parameters of the spatial filter such as the pitch and number of grating lines are temporally variable, even during measurements,
3. For a single input image, multiple channels of identical or different spatial filters can be arranged in parallel or simultaneously.

All these flexible features come from the fact that the patterns are generated electronically. Some examples using these advantages are briefly introduced here.

a Practical SFV systems usually employ rectangular transmittance for the spatial filter, but sinusoidal transmittance is ideal because no higher harmonic component appears, as discussed in Sect. 2.3. The electric spatial filter can easily realize a sinusoidal pattern, although it is discrete in the pixel unit. A negative transmittance is also easily made, so that the pedestal component can be eliminated.

b A complex logarithmic mapping of input moving images transforms their scaling and rotating motions to translating ones [120]. Then, these motions can be measured by the spatial filter introduced in the transformed domain [121]. For simplicity of image processing, it should be better to multiply the input image by the weighting function in the input image domain. The weighting function in the transformed domain is inversely transformed into the input image domain and the input image is multiplied by it pixel by pixel. It is generally difficult to construct an actual spatial filter having such an inversely transformed weighting function, but it is possible with the electric spatial filtering system.

c Since the spatial filter picks up the specific spatial frequency component of $1/p$ from the intensity pattern of a moving image being measured, the output signal amplitude depends on the spectral distribution of spatial frequency components contained in the image pattern. The example of Fig. 5.43 includes the pitch adjustment circuit [119] which analyzes the spatial frequency spectrum of the input image, searches the spectrum peak, calculates the optimum grating pitch, and then sends the control signal to adjust the pitch to the pattern generator. To evaluate the adaptiveness of the adjusted grating pitch to the input image and follow its change, the power of output signals

$$\left| F\left(\frac{1}{p}\right) \right|^2 = \langle g_{\mathrm{c}}^2\left(t\right) + g_{\mathrm{s}}^2\left(t\right) \rangle \tag{5.27}$$

is monitored and returned as feedback signals into the pitch adjustment circuit. In (5.27), $\langle \cdots \rangle$ denotes a temporal average. The above operation can be done at the same time with the frequency measurements for the velocity determination. If necessary, two or more different signal processings can be performed simultaneously for a single input image.

d The moving grating, which is used in the frequency-shifting technique for directional discrimination, is also realized by the generation of a translation or time-varying grating pattern in the pattern generator. Though this is a possible approach, the example of Fig. 5.43 enables us to take a different approach to frequency shifting. When the grating is moving at velocity v_{g} in the x direction, the expected output signal is expressed by

$$g_{\mathrm{M}}\left(t\right) = \iint f\left(v_{\mathrm{x}}t - x, v_{\mathrm{y}}t - y\right) \cos\left[\frac{2\pi\left(x - v_{\mathrm{g}}t\right)}{p}\right]\,\mathrm{d}x\,\mathrm{d}y\ . \tag{5.28}$$

By using (5.25) and (5.26), this equation is rewritten as

$$g_{\mathrm{M}}\left(t\right) = g_{\mathrm{c}}\left(t\right)\cos\left(\frac{2\pi v_{\mathrm{g}}t}{p}\right) + g_{\mathrm{s}}\left(t\right)\sin\left(\frac{2\pi v_{\mathrm{g}}t}{p}\right)\ . \tag{5.29}$$

By using the second generator that gives two sinusoidal signals $\cos(2\pi v_{\mathrm{g}}t/p)$ and $\sin(2\pi v_{\mathrm{g}}t/p)$, multiplication in the first and second terms of (5.29) is done for every frame, as shown in Fig. 5.44 [116, 117]. Hence, the same effect as that with the moving grating pattern is obtained by the simple multiplication of signals in the frame rate instead of multiplication of the input pattern by the moving grating pattern pixel by pixel.

e Two-dimensional velocity components can be measured by generating two grating patterns that are orthogonal in the directions of the grating lines. If the moving direction is discriminated for each velocity component in terms of the above-mentioned approaches, a two-dimensional velocity vector is determined.

The electronic spatial filtering system depicted in Fig. 5.43 yields two periodic signals $g_{\mathrm{c}}(t)$ and $g_{\mathrm{s}}(t)$ given in (5.25) and (5.26), respectively, having the same central frequency with the phase difference of $\pi/2$. The complex signal that contains $g_{\mathrm{c}}(t)$ and $g_{\mathrm{s}}(t)$ as its real and imaginary parts, respectively, belongs to an analytic signal [66, 75], and $g_{\mathrm{c}}(t)$ and $g_{\mathrm{s}}(t)$ are a Hilbert

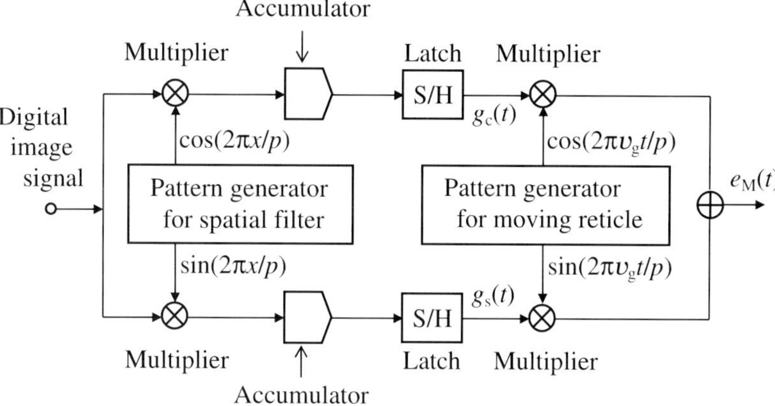

Fig. 5.44. Use of a signal generator for directional discrimination in an electronic spatial filter [116, 117]

transform pair. By using the one-dimensional Fourier spectrum $F(\mu)$ of the image intensity $f(x, y)$ in the x direction, the two signals $g_c(t)$ and $g_s(t)$ are expressed as

$$g_c(t) = \left| F\left(\frac{1}{p}\right) \right| \cos \left[\frac{2\pi x}{p} - \phi \left(\frac{1}{p}\right) \right] , \tag{5.30}$$

$$g_s(t) = \left| F\left(\frac{1}{p}\right) \right| \sin \left[\frac{2\pi x}{p} - \phi \left(\frac{1}{p}\right) \right] , \tag{5.31}$$

where

$$\phi \left(\frac{1}{p}\right) = \phi_{1/p}(t) = \tan^{-1} \frac{g_s(t)}{g_c(t)} , \tag{5.32}$$

which denotes the instantaneous phase $\phi(t)$ of the image at the spatial frequency $\mu = 1/p$ and at time t. Since the time rate of a phase change of the analytic signal gives an instantaneous frequency $f_0(t)$ as

$$\begin{aligned} f_0(t) &= \frac{1}{2\pi} \cdot \frac{d}{dt} \left[\tan^{-1} \frac{g_s(t)}{g_c(t)} \right] \\ &= \frac{g_c(t) \cdot g_s'(t) - g_c'(t) \cdot g_s(t)}{2\pi \left[g_c^2(t) + g_s^2(t) \right]} , \end{aligned} \tag{5.33}$$

where $g_c'(t)$ and $g_s'(t)$ are the time derivatives of $g_c(t)$ and $g_s(t)$, respectively. Without using spectral or correlation analysis, then, the instantaneous value of the central frequency of the signal can be estimated. The instantaneous amplitude or envelope of the analytic signal is given by

$$a(t) = \sqrt{g_c^2(t) + g_s^2(t)} , \tag{5.34}$$

and the temporal average of the square of $a(t)$ gives the mean power described in (5.27). A change of 180° in the moving direction of the image yields a reversal of the phase rotation and a change in the sign of the instantaneous frequency estimated by (5.33). Thus, directional discrimination can be achieved by monitoring the sign of $f_0(t)$. Equation (5.33) indicates that an estimation of the instantaneous frequency requires numerical differentiation of $g_c(t)$ and $g_s(t)$, which may cause systematic errors due to frame rate sampling. To improve this problem, the use of a discrete approximation has also been studied [121]. This method has been proposed in the framework of the present type of spatial filter. The temporal resolution of the electronic spatial filter is limited to the frame rate and, thus, usual methods for frequency measurements such as FFT or correlation analysis are inappropriate because they require sufficient integration of data for precise measurements. Instantaneous determination of the central frequency is effective for frame-rate measurements.

Instantaneous frequency measurements of this type have also been extensively studied and actively employed [62] in a spatial filtering system using the CCD line sensor which is described in Sect. 5.7.2. The SFV system shown in Fig. 5.42 is able to produce two phase-orthogonal output signals, which are used for determination of the instantaneous frequency based on the same principle mentioned above. This signal processing method may be referred to as a phase angle measuring technique [122] and is effective for measurements of velocity fluctuations and those with high temporal resolution [62, 114]. An example of the discrete approximation is simply described as [123]

$$\Delta\phi_{\mathrm{m}} = \tan^{-1}\frac{g_{\mathrm{c}}\left(m\right)g_{\mathrm{s}}\left(m-1\right) - g_{\mathrm{c}}\left(m-1\right)g_{\mathrm{s}}\left(m\right)}{g_{\mathrm{c}}\left(m\right)g_{\mathrm{c}}\left(m-1\right) - g_{\mathrm{s}}\left(m\right)g_{\mathrm{s}}\left(m-1\right)}, \tag{5.35}$$

where $\Delta\phi_{\mathrm{m}}$ is the phase difference between two phase angles at the mth and $(m-1)$th samplings and $g_{\mathrm{c}}(m)$ and $g_{\mathrm{s}}(m)$ are instantaneous amplitudes at the mth sampling of $g_{\mathrm{c}}(t)$ and $g_{\mathrm{s}}(t)$, respectively. By using the sampling interval Δt, the time rate of the phase change is expressed as

$$\frac{\mathrm{d}}{\mathrm{d}t}\left[\tan^{-1}\frac{g_{\mathrm{s}}\left(t\right)}{g_{\mathrm{c}}\left(t\right)}\right] \cong \frac{\Delta\phi_{\mathrm{m}}}{\Delta t} \tag{5.36}$$

and, then, the estimated central frequency $f_0(m)$ at the mth sampling is expressed by

$$f_0\left(t\right) \cong f_0\left(m\right) = \frac{1}{2\pi} \cdot \frac{\Delta\phi_{\mathrm{m}}}{\Delta t}. \tag{5.37}$$

Since the instantaneous phase and frequency are calculated directly from sampled values of $g_{\mathrm{c}}(t)$ and $g_{\mathrm{s}}(t)$, they are sensitive to noise and signal quality. In flow measurements, for example, the measuring volume is almost always passed by more than one particle, and the output signal may include phase discontinuity or jump and irregularity in a cycle. The resultant phase and frequency contain fluctuations and, thus, a statistical treatment may be necessary, as discussed in the literature [62]. To increase the measurement accuracy, an averaging algorithm using squared amplitudes has been implemented as [123]

$$f_0\left(m\right) = \frac{1}{2\pi\,\Delta t} \cdot \frac{\sum_{m=1}^{q}\left[g_{\mathrm{s}}^2\left(m\right) + g_{\mathrm{c}}^2\left(m\right)\right]\Delta\phi_{\mathrm{m}}}{\sum_{m=1}^{q}\left[g_{\mathrm{s}}^2\left(m\right) + g_{\mathrm{c}}^2\left(m\right)\right]}\,, \tag{5.38}$$

where q is the number of samples for averaging.

5.8.2 Computer Image Processing

The electronic spatial filter in the above subsection processes the sensor signal using exclusive electronic hardware. This method is used for real-time and on-line measurements of relatively low velocity. If off-line treatments are accepted or rather low velocity is of interest, video signals are processed fully by computer software to realize the spatial filtering effect [124–128]. Although processing time should be considered due to the software computation, the flexibility of data analysis is an attractive feature. In addition, the method requires only a camera system and a desktop computer with appropriate software, but no specific optical nor electronic device.

Basic signal processing in the method is equivalent to that used in the electronic spatial filtering method in Sect. 5.7.1. Analog video signals from the camera system are captured frame by frame and digitized into sequential pixel images in the image-capture board/card on the computer. Direct digital outputs from the digital camera system are more favorably used in the computer with better image quality. Consider a sequential scene with pixel images $f(v_{\mathrm{x}}t - x, v_{\mathrm{y}}t - y)$ of N frames in the pixel-coordinate (x, y) at time t (or at frame number t). A sinusoidal transmittance pattern $\cos(2\pi x/p)$ of the gray level, where p is the grating pitch, is generated by the software and stored in the computer before starting the following processing. As shown in Fig. 5.45, the software superposes the sinusoidal pattern on the image of respective frames, and, then, the resultant image is expressed as

$$I_{\mathrm{t}}\left(x, y\right) = f\left(v_{\mathrm{x}}t - x, v_{\mathrm{y}}t - y\right)\cos\left(\frac{2\pi x}{p}\right). \tag{5.39}$$

A temporal output $A(t)$ at time t is obtained by accumulating the gray value in the image of frame number t as

$$A\left(t\right) = \sum_{x}\sum_{y} I_{\mathrm{t}}\left(x, y\right). \tag{5.40}$$

Then, the sequential output signal $A(t)$ yields the results of calculating N frames. Note that (5.40) with (5.39) is equivalent to the expression of (5.25). According to the principle of spatial filtering velocimetry, the amplitude of

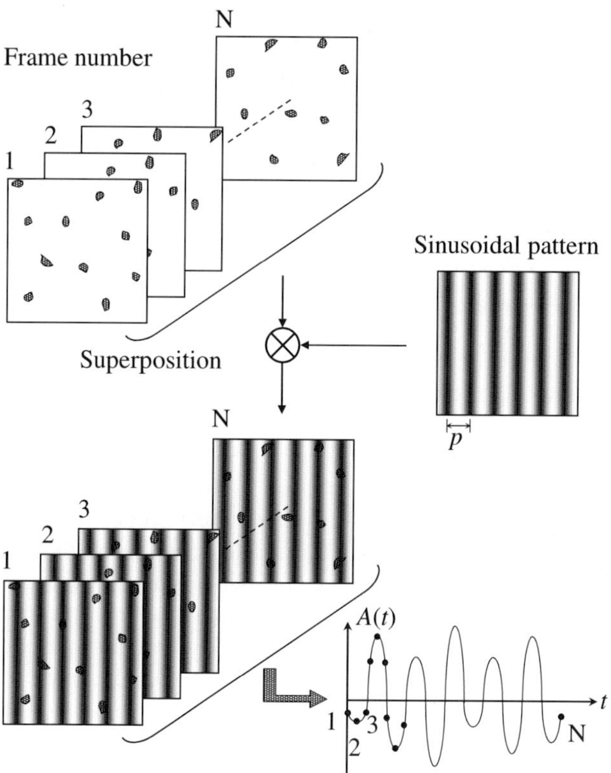

Fig. 5.45. Principle of the image processing for electronic spatial filtering velocimetry

$A(t)$ is periodically modulated and its central frequency is proportional to the velocity component v_x. Thus, the object's velocity is determined by measuring the central frequency with appropriate signal processing techniques such as spectral analysis, as described in Chap. 4.

Since the periodic transmittance pattern is generated by the software operation in the computer, a change in the pattern is easily made, for example, changes in the grating pitch and direction and even a choice of the transmittance function from sinusoidal and rectangular patterns. As discussed in Sect. 2.6, the grating pitch should be determined adaptively to the feature of moving images, especially the image size of scattering objects or particles and the correlation length of random image-intensity patterns. In the off-line analysis, then, the correlation length of video images is calculated and an optimum value of the grating pitch is chosen [127]. During measurements, the grating pitch is varied in accordance with a change in the correlation length. Therefore, better measurement accuracy is expected by this adaptive operation than that using a fixed value for the grating pitch. If the two velocity components v_x and v_y be measured, two-channel parallel processings are used with two transmittance patterns for the x and y directions. This is available for off-line analysis of video images stored in memory media, but may be disadvantageous in real-time performance due to the increase in computation time. The effect of a moving grating is also obtained by generating a translational pattern of the periodic transmittance and is used for discrimination of the velocity direction on the basis of frequency shifting.

To obtain better performance in dynamic image processing, a computer with faster CPU and larger memory capacity is generally desirable. With a desktop personal computer, the number of frames which can be processed may be small due to the size of the memory, and, then, the sequential output signal $A(t)$ may consist of short data records. This means that the recorded signal $A(t)$ contains insufficient periodic cycles by which the central frequency could not be precisely measured by spectral or correlation analysis. In this case, the maximum entropy method (MEM) (see Sect. 4.2.4) is a better choice for spectral analysis. Usually, the MEM gives more realistic frequency estimates with better resolution than other conventional methods such as the FFT for short data records [128].

The present computer-based spatial filtering method provides another useful application, that is, simulation studies of SFV. When video images of moving objects being measured are available, generation and processing of SFV signals and the measurement procedure are simulated by the above-mentioned software operation in the computer. If such a simulation is repeated with different parameters of the spatial filter and different signal processing methods, the optimum condition is estimated for the spatial filter and signal processing. Thus, this is used for optimum design of the SFV system without conducting actual measurements by trial and error. In addition, if the moving image of objects is also generated by computer simulation, the whole process of SFV measurements is simulated within the computer for the object of interest [128]. Hence, various evaluations are made in advance for practical SFV measurements, such as the optimum parameters of the spatial filter, optimum signal processing, measurement accuracy, signal properties, and other problems and solutions.

5.9 Survey of Spatial Filtering Devices

Several types of the device and system can be used for spatial filtering velocimetry, as described in this chapter. In general, the optimum type is chosen by considering measurement circumstances such as image intensity, contrast and quality, size and concentration of scattering objects, time- and space-varying characteristics of objects, environmental condition, velocity range of object movement, scale of probing volume, and so on. Since each type of SFV device and system has advantages and disadvantages, it is useful to survey briefly their features in the last section of this chapter.

Table 5.1. Spatial filtering devices and their usages for typical measurement operations

Device	Type	Pedestal removal	Directional sensing	Two-dimensional measurement	Remarks
Transmission grating	Parallel slits (basic)	2 imaging planes with gratings	FS[c]: moving or rotating grating; PS[d]: 2 gratings with $\pi/2$	2 imaging planes with gratings	General purpose; low cost
Prism grating	Beam deflector	2 stages	PS: 3 stages	Piled gratings	Simple optical system
Lenticular grating	Beam deflector	2 deflection angles	FS or PS: 4 deflection angles	Piled gratings	Simple optical system; low cost
Optical fiber array	Light guide	2 fiber bundles	FS or PS: 4 fiber bundles	2 optical imaging systems	Flexible optical system
Liquid crystal cell array	Parallel slits	Electric HPF[a]	FS: moving grating	Piled arrays	Grating flexibility
Solar cell array	Detector	2 detection channels	PS: 4 detection channels	2 optical imaging systems	Simple optics
Line sensor	Detector	2-channel DMP[b]; weighting operation	FS or PS: 4-channel DMP	2 optical imaging systems	Grating flexibility; simple optics
Area sensor	Imaging (camera)	Weighting operation	FS or PS: phase-orthogonal operation; FS: time-shifting operation	2 weighting operations	Grating flexibility; simple optics; object monitoring; off-line monitoring

[a] High pass filter   [b] Demultiplexing   [c] Frequency shifting   [d] Phase shifting

The transmission grating is a basic, simple, and low-cost device for SFV. Since the grating includes no additional operation in itself, the SFV system using this device needs consideration in the optical and signal-processing systems for development of measuring functions: pedestal removal, directional discrimination, and two-dimensional velocity measurements. From a different point of view, however, the simplicity of the transmission grating is suitable for wide use and general purposes. The prism grating serves as both a beam-splitting optical element and a periodic grating and, thus, is advantageous for differential detection to remove a pedestal component. As this grating is not a very popular optical element in the market, only limited ranges of grating pitch and size are available. The wavelength dispersion of the prism material should also be noticed if illumination by white light is used. The lenticular grating is an inexpensive optical element which works as both a beam splitter and a periodic grating. It is particularly useful for the configuration of more than two detection channels, which are effective for directional discrimination as well as pedestal removal. The optical fiber array is a light-propagating optical element which serves simultaneously as both a beam splitter and a periodic grating. Owing to the use of fibers, flexibility in the optical system comes particularly between the imaging system and the photodetector with signal processing units. This is advantageous for measurements under severe environmental conditions. The fiber array is a homemade or custom-made optical device and is inconvenient for general purposes. The liquid crystal cell array is also a special device which must be designed and fabricated exclusively for SFV measurements. Since it is a time- and space-variable transmission grating driven by voltage application, the flexibility of the grating pitch and shifting velocity is a main advantage. Low-frequency response and low contrast of the LC cell array are problems that must be addressed for practical uses.

Although the above-mentioned five types of SFV devices need photodetectors for a system, the detector and imaging types are photodetectors themselves and contribute to simplification of the optical system. For measurements in an extremely low light level, specific detectors such as photomultipliers are necessary, but the detector and imaging types are almost unavailable. The solar cell array is a simple and compact SFV device which easily realizes differential detection to eliminate the pedestal component, but its sensitivity is insufficient for week-intensity images. The photodiode array, CCD linear sensor, and area sensor are very favorable for SFV devices, since they are commercially available and quite popular recently, and also their outputs are conformable to digital electronics and computer. As a variety of measuring functions are realized in electronic operations, the grating pattern including its parameters is variable and such a flexibility is an attractive feature. The one-dimensional type of device is advantageous for measurements of a single velocity component, whereas two-dimensional velocity components should be treated by the imaging type or area sensor. Another practical advantage of the area sensor is its facility for monitoring the object during measurements. Computer-based SFV systems using video images are very effective for off-line measurements and repeatable analysis, but frame-rate limitation should be considered. A list of SFV devices treated in this chapter is given in Table 5.1 in relation to the methods for realizing the three measuring functions.

6

Applications

Spatial filtering velocimetry (SFV) is a practical tool and provides a measuring system that can be compact, inexpensive, and easy to handle. It gives the same order of accuracy as that obtained with laser Doppler velocimetry (LDV). These features are due to its simplicity in measuring principle and configuration. The LDV technique is well known for its high spatial resolution, but this feature inevitably involves the fact that the probe region is small by focused laser beams and a larger macroscopic probe area cannot be covered by its normal use. The SFV technique covers a wide range of scales of motion, for example, from fluid flows in a microscopic region to the ground observed from an airplane. Without difficulty, this is done by using the optical imaging capability together with the spatial filter window and illuminating condition. This scaling flexibility for objects being measured promotes applications of the SFV technique to various fields of science, engineering, and biomedicine. The principle of SFV is also used for measuring the length, distance, displacement, and other quantities related to the velocity and for detecting the focus of moving objects as a result of image formation. There are also other techniques for velocity measurement that are similar or related to the SFV technique.

In this chapter, various examples of applications of the SFV technique are reviewed. Before learning the examples, the fundamental performance of the SFV system is discussed in Sect. 6.1. Sections 6.2–6.4 are devoted to several examples of SFV applications. SFV measurements except for the velocity sensing are next presented briefly in Sect. 6.5. In Sect. 6.6, some related velocimetric techniques that are based on a principle similar to that of the SFV technique are described. Finally, Sect. 6.7 gives a brief comparison of SFV and LDV techniques.

6.1 Performance

Although there are several aspects which describe the performance of a total SFV system, the principal measures are accuracy, linearity, and resolution.

They are influenced mainly by the optical system and the signal-processing system. Note that, very often, the conditions of moving objects themselves and measurement circumstances are also the factors which must be considered in performance.

6.1.1 Accuracy

As described in Sect. 2.4.1, the fundamental accuracy of central frequency and hence velocity measurements is determined by the number of grating lines in the spatial filter. This number determines the maximum number of cycles contained in the output signal of one burst from a particle passing through the probe volume and, thus, the minimum width of a spectral peak at the central frequency. This type of spectral broadening is equivalent to transit time broadening, which is sometimes called also "ambiguity" broadening in laser Doppler velocimetry [5]. In usual measurements, spectral broadening is caused by different sources. The main sources of broadening in the spectra of SFV signals are as follows:

1. number of grating lines
2. random fluctuation
3. defocus and aberration
4. velocity distribution in space
5. velocity variation in time.

The number of grating lines is the most elementary source of spectral broadening and the effect is inevitable in SFV measurements. Broadening from other sources is added to elementary broadening, and accuracy is further degraded. A basic approach is, thus, to have as large number of grating lines as possible in the system to make elementary broadening smaller. To follow this, a smaller grating pitch and a larger window are desired in a spatial filter. There is a lower limit for grating pitch owing to a signal visibility (see Sect. 2.6.2) and geometric size. A large window for a spatial filter takes a risk of including the velocity distribution as will be discussed later. Hence, a large filter window makes elementary broadening small, but broadening due to the range of velocities comes in.

If a moving object is spatially continuous or presented for a sufficiently long time (not like a small particle) in the probe volume without intermittence, the oscillatory output may continue for a large number of cycles, even for a small number of grating lines. Then, no spectral broadening might be expected. However, such an output normally has randomness in the amplitude and phase of oscillations, as shown in Fig. 4.1a. As described in Sect. 4.1, this type of random fluctuation statistically contains a limited temporal correlation length which is related to transit time τ_T or to the time for passage of a single particle through the probe volume. Then, the number of cycles included in the transit time, equivalent to one burst period, in the output makes sense of the spectral width. As a result, the signal spectrum does not show a perfectly defined frequency but a distribution centered at the central frequency. The broadening in this mechanism is sometimes called "random phase broadening" [5] and, after all, it is due to the number of grating lines.

Errors in image formation often arise from defocus and aberration. Even in an imaging system with well-defined magnification, defocus is caused by the axial displacement of a moving object, owing to fluctuating motion or the velocity component along the imaging optical axis. Use of an imaging lens with a small diameter causes some aberrations which cannot be neglected. In the presence of defocus and/or aberration, the image of a moving object is blurred on the plane of the spatial filter. This results in the loss of contrast and sharpness of the image, causing a decrease in the SNR and preventing measurements from being accurate. The distortion impinges on the linearity between the object's velocity and the image velocity (see Fig. 3.7), directly causing errors in the measured velocity.

The broadening mechanisms 1–3 are basically due to conditions in the SFV optical system. Even for an optical system with ideal conditions, spectral broadening occurs as an effect of the moving condition of an object, resulting from variations of velocity in space and time. It is very usual in measurements of the flow velocity in a fluid. In a tube or pipe, the flow has a gradient of velocity, and a distribution of velocities of suspended particles is contained in the probe cross-sectional area. A range of velocities produces different central frequencies in SFV signals and gives spectral broadening. In other words, different velocities result in different periods of cycles in signals for the passage of particles through a probe volume. Then the phase relations of these cycles are not maintained but change in a shorter period than the transit time. Consequently, the number of cycles included in one burst in which the phase relation is maintained decreases and the spectrum broadens. Of course, the use of a smaller probe volume reduces the effect of the spatial velocity distribution.

A temporal variation of velocity also produces different central frequencies during a measurement period, and the broadening of the spectrum occurs in the same way in the presence of the spatial velocity distribution. The flow velocity of a fluid is often applicable to this case because of turbulence or pulsation. The problem of velocity variations with time can be solved by signal processing techniques which treat real time or time-resolved measurements. However, note that such techniques are often sensitive to noises which are another source of lowering accuracy since they are not the technique that reduces noises by time integration.

Finally, it may be mentioned that measurement accuracy is influenced by electric noises in the photodetector and the signal processing system. This is a common source of trouble in general electronic instruments. As described in Chap. 4, signal processors that are based on time integration such as spectrum analysis and correlation measurement reduce the effect of electric noises on accuracy. The trouble becomes serious with the use of the counting technique and the phase angle measuring technique, since they process the signal amplitude directly.

6.1.2 Linearity

Linearity means a proportional relation between an object's velocity and the measured central frequency of the output signals, which is theoretically given by (1.1). The SFV optical system puts no limit on linearity with the accuracy mentioned above, but the signal processing system including the photodetector does not guarantee it in higher signal frequencies because of the system bandwidth. If an expected central frequency or a given velocity exceeds the upper cutoff frequency of the processor's passband, the signal power at the frequency is substantially attenuated. In this frequency range, the real central frequency cannot be determined correctly but the estimated central frequency may be lowered due to the decreasing frequency response of the system in the cutoff range. Hence, the estimated value of the central frequency is saturated as velocity increases and the relation of frequency to velocity becomes non-linear. The system bandwidth is determined by the total frequency response of the photodetector, amplifier, and other electronic instruments employed in the signal processing system. The dynamic range of the SFV system for velocity is therefore defined as the linear extent of the frequency response up to the cutoff frequency of the total electronic system.

6.1.3 Resolution

In using spatial filtering devices provided by optical elements such as the transmission grating, prism grating, lenticular grating, and fiber array, temporal resolution is determined mainly by the signal processing system. Frequency tracking and counting techniques give the high resolution so that they can be better used for real time measurements than with the time-integration type processing such as spectrum and correlation analysis. Statistical analysis includes averaging, and the histogram, and probability density requires a substantial measurement period for acquiring sufficient data and, thus, is usually not advantageous for temporal resolution. When the spatial filtering device employed is of a type with electronic control such as a liquid crystal cell array and 1-D or 2-D detector arrays, the temporal resolution is limited by the maximum control frequency of SFV devices, rather than by the bandwidth of a signal processing system.

Spatial resolution is determined by the sizes of the probe volume in both the cross-sectional plane and the axial direction. To increase the spatial resolution, the size of a spatial filter window should generally be reduced and the imaging magnification enlarged while the required number of grating lines is maintained. For high axial resolution, side illumination and a small focusing depth are desirable. Note that one of the advantages of the SFV technique is covering probe section of a wide area such as a river, road, or ground, although it is opposite in nature to spatial resolution.

6.2 Measurements of Flow Velocity

One of the most interesting applications is measurements of flow velocity and its profile across a section of fluid. Optical access is advantageous for studies of flows because a light beam does not disturb the flow and can be used as a micron-size probe when focused. For this application, the LDV technique is well known as a useful tool, but LDV instruments are generally expensive and need accurate optical alignment. In many cases of flow measurements, SFV can replace LDV. The flow velocity is obtained from measurements of small scattering particles suspended in the fluid. Thus, it is assumed that particles follow the fluid flow sufficiently .

6.2.1 Transmission Grating Velocimeter for a Microscopic Region

To measure the flow velocity in a microscopic region using the SFV technique, laser light should be used as an illuminating source since it is advantageous for focusing the light into a microscopic probe volume with sufficient intensity. Imaging of small particles is done by a microscope with the necessary magnification. In a case including such optical parts, the measuring system should be constructed in a practical form so that it guarantees simple alignment and mechanical stability. Figures 6.1 and 6.2 show a schematic diagram and a photograph of a differential-type transmission grating velocimeter for a microscopic region [28]. The basic construction follows the optical system shown in Fig. 5.5. The illumination system is separated from the detection system, and the angle between their optical axes is determined in accordance with the object being measured, for example, by considering whether it is transmissive or reflective. For flow measurements in usual cases, an angle of

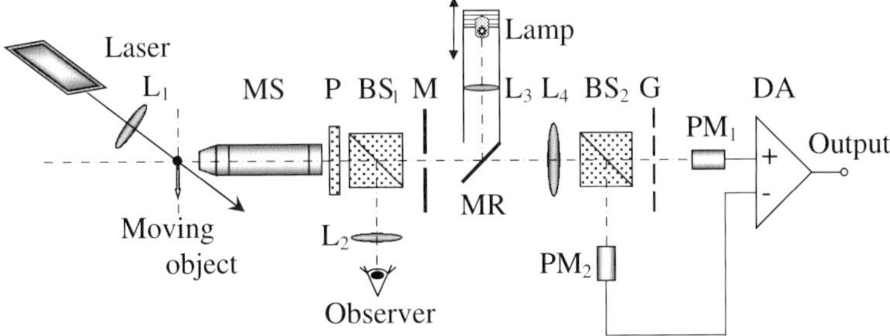

Fig. 6.1. Configuration of the differential-type transmission grating velocimeter for a microscopic region

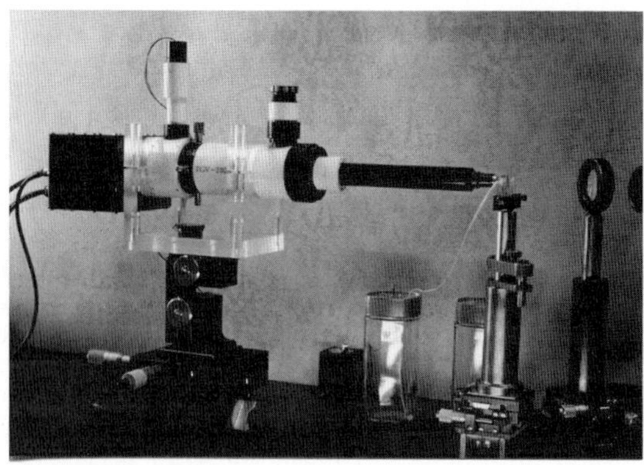

Fig. 6.2. Photograph of a differential-type transmission grating velocimeter for a microscopic region

around 30° may be convenient since it gives a near-forward scattering scheme under dark-field illumination.

In the detection system, the light scattered by particles is received by a microscope (MS). A polarizer (P) is used to reduce the undesirable ambient or background light and improve the image contrast. A beam splitter BS_1 is included to allow visual observation, which is necessary for focusing the image and positioning the probing area. A small aperture of mask M defines the probe cross-sectional area. When the position of this aperture must be known on the object image, a lamp assembly is inserted between the mask M and a lens L_4, as shown in Fig. 6.1. The aperture is illuminated by this lamp and imaged back onto the probe cross-sectional area in the object plane as a small bright spot, which can be observed via the beam splitter BS_1 by the observation system. The lamp assembly is taken off while measurements are being made. The received light is finally detected by photomultipliers. An extremely large magnification is not available for imaging in the detection system because of the limit in the working distance of the objective being used. The upper limit of the imaging magnification is $\sim$200 X for this system. This limit determines the spatial resolution of $\sim$5 μm when a 1-mm diameter aperture is used. The total detection system is mounted on XYZ translation stage with manipulators, as shown in Fig. 6.2. The system is rotatable around the detecting optical axis so that the grating lines can be set perpendicularly to the flow direction with an eyepiece of the observation system.

Figure 6.3 [25] demonstrates measured frequency histograms of the flow velocity distribution of water along the cross-sectional axis of a cylindrical glass tube having a 700-μm diameter. By using two different imaging magnifications $M = 14.4$ and 36 with an aperture of 1-mm diameter, the probe area was set at about 70 and 28 μm in diameter in figures (**a**) and (**b**), respectively. The light source employed was a 10-mW He-Ne laser. By using a Ronchi grating having a line interval of $p = 254$ μm as the spatial filter, the number n of grat-

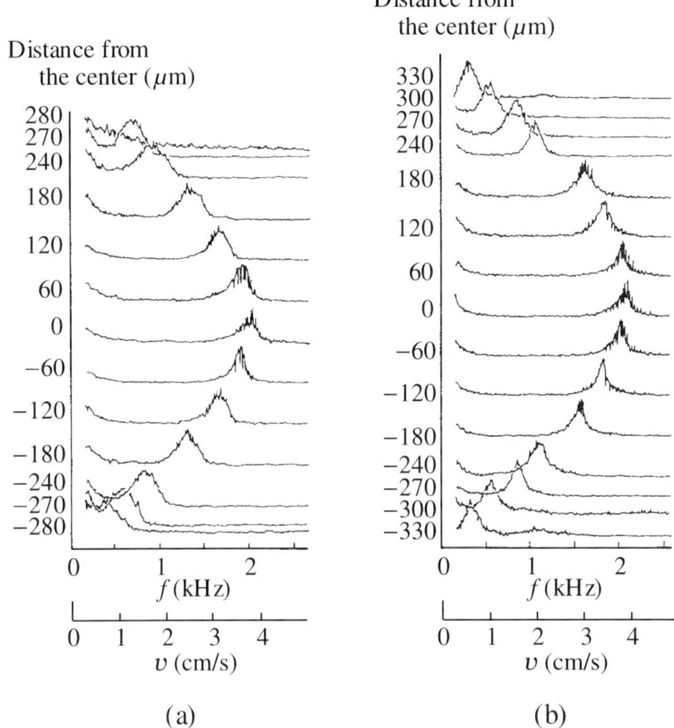

Fig. 6.3. Typical frequency histograms of the flow velocity along the cross-sectional axis of a cylindrical glass tube having a 700-μm diameter. The probe area was (**a**) 70 and (**b**) 28 μm in diameter [25]

ing lines was set at 4. Despite such a small number, a flow velocity distribution was successfully obtained, giving good agreement between theory (assuming a Newtonian flow) and measurements [25]. A large probe volume makes an error due to the range of velocities in the distribution, which is especially noticeable near the wall of the tube. Comparison of Figs. 6.3a and b indicates that result (**b**) for the 28-μm diam probe volume gives better frequency histograms than result (**a**) for the 70-μm diam probe volume at locations outside $\pm 240\mu$m from the tube center. Thus, the size of the probe volume is particularly important for measurements of the flow velocity distribution. Attention should also be paid to strong scattered light from the wall, which may degrade the signal-to-noise ratio. For measurements involving a cylindrical glass tube, note that the tube having fluid inside has the effect of a cylindrical lens on the illumination and detection geometry. To avoid this effect, the tube can be immersed in a fluid-filled glass box having a plain surface.

Ushizaka and Asakura [23] carried out a comparison study of SFV with LDV techniques for measurements of the flow velocity distribution in various small glass tubes having cylindrical and rectangular cross sections. They observed that comparable accuracy could be obtained in both techniques. Water usually contains a sufficient amount of naturally occurring particles for measurements. If the number of particles available for scattering is small, a very small amount of scatterers such as milk, acrylic particles, or polystyrene particles may be seeded into the water to increase the particle concentration. However, one should be careful of the concentration since a large number of particles in the probe volume causes intensity summation of many periodic waveforms in random phases in the photodetector output, and this can be a source of signal deterioration and an increase in shot noise. The effect of particle concentration on spectral broadening was theoretically and experimentally studied [27], and a modified signal processing method was proposed to improve the measurements in large particle concentrations [28]. The phase-shifting technique, which is used for eliminating directional ambiguity, as described in Sect. 5.1.4, is also influenced by large particle concentration, since the signal phases are not consistent in random addition and the desired phase difference of $\pi/2$ is not always maintained [26]. A modified signal processing method was successfully employed to reduce the effect of large particle concentration on measuring the temporal variation of flow velocity including a reversal of the flow direction [28].

Electrophoretic mobility distributions of suspended particles have also been measured [31] by a transmission grating velocimeter. To achieve high velocity resolution, specially designed optics consisting of a dark-field condenser and objective lenses and a rectangular vertical electrophoresis cuvette were employed. The probe volume was set at 0.8×0.5-mm^2 cross section with a 10–20-μm depth positioned in the middle plane of the cuvette where the velocity gradient is almost zero. Under the application of an electric current I, the mean velocity v of diluted erythrocytes (red blood cells) suspensions was determined by using the SFV principle and, then, the mean electrophoretic mobility $\bar{\mu}_{25}$ was deduced from the relation

$$\bar{\mu}_{25} = \sigma_{25}\frac{S}{I}v \,, \tag{6.1}$$

where σ_{25} denotes the conductivity of suspensions at 25°C which was precisely measured separately, and S is the cross section ($1 \times 12\,\mathrm{mm}^2$) of the cuvette used. The experimental results demonstrate a narrow frequency peak in the power spectrum which achieves the specific bandwidth, defined by (2.38), of $D = 0.01$, and high velocity resolution was realized in their measurements. This performance was significantly better than with the LDV technique. Such a remarkable resolution was due mainly to the use of a well-designed dark-field optical system and a rectangular (not cylindrical) cuvette with reasonable dimensions, together with the number of grating lines, $n = 100$.

6.2.2 Two-Dimensional Vector Velocimeter

Two-dimensional vector components of a flow velocity distribution have been measured by the frequency-shifting technique using a rotating disk with dual ring gratings, which was already described in Sect. 5.1.5. The image of particles suspended in the fluid flow is first formed on the mask M in the same way as that in Fig. 6.1. By means of a relay lens, beam splitter, and prisms, two identical images of particles spatially restricted by the mask M are formed on gratings G_x and G_y. The configuration of the disk gratings is described in Fig. 5.18 [106]. Alternatively, an optical system using an image fiber bundle has also been reported and is shown in Fig. 6.4 [105]. The fiber bundle allows the object and detecting fields to be set separately in different places. This flexibility is advantageous for applications since rotation of the grating disk by an electric motor may yield mechanical vibration which is unwelcome in the microscopic probing in the object field. The bundle usually consists of a periodic arrangement of several thousands of fiber elements, which causes another undesirable spatial filtering effect on output signals [105, 106]. The effect can be eliminated by using differential detection in the same way as pedestal removal.

As described in Sect. 5.1.5, the central frequencies $f_{x'}$ and $f_{y'}$ of two detector outputs vary according to the magnitude and direction of the image velocity under given shift frequencies $f_{gx'}$ and $f_{gy'}$ of gratings G_x and G_y. Figure 6.5 [106] shows power spectral distributions of two velocity components $v_{ix'}$ and $v_{iy'}$ along the cross-sectional axis of a cylindrical glass tube having a 3.4-mm diameter, for four different angles θ' of the flow direction with respect to the x' axis or the moving direction of grating lines in G_x. The diameter of the probe cross-sectional area was set at about 110 µm, and the

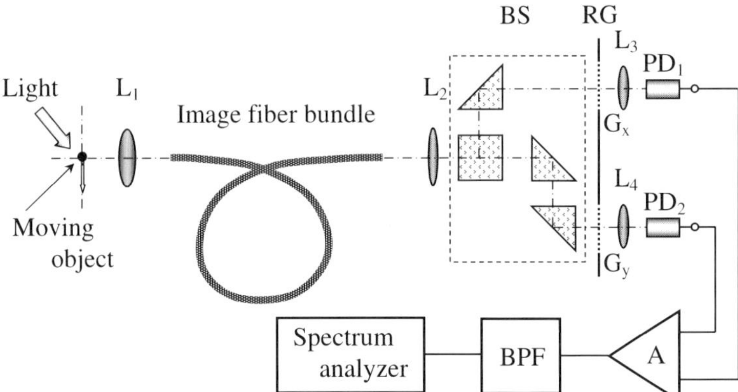

Fig. 6.4. Transmission grating velocimeter using an image fiber bundle for measuring two-dimensional velocity components (from [105], Masson 1989)

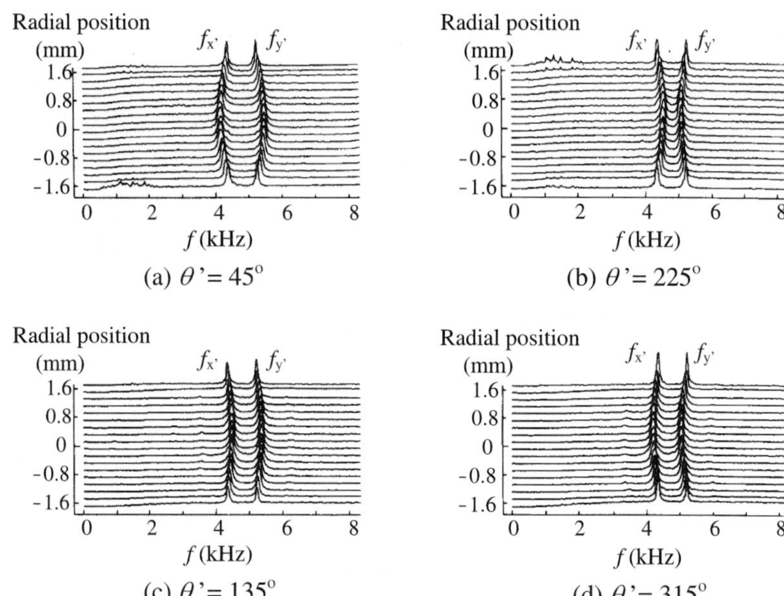

(a) $\theta' = 45°$

(b) $\theta' = 225°$

(c) $\theta' = 135°$

(d) $\theta' = 315°$

Fig. 6.5. Typical power spectra of two velocity components $v_{ix'}$ and $v_{iy'}$ along the cross-sectional axis of a cylindrical glass tube having a 3.4-mm diameter [106]

shift frequencies $f_{gx'}$ and $f_{gy'}$ were fixed at 4.3 and 5.175 kHz, respectively. The results demonstrate that the central frequency of the velocity component is shifted higher or lower from each shifting frequency due to a change in the flow direction relative to the grating movement. The flow velocity distribution and direction can be obtained from these spectra according to the relations (5.10) and (5.11). Errors in measurements with the rotating disk gratings are usually caused by inhomogeneity of the pitch of grating lines within the circular area restricted by the mask image. To reduce the errors, the ratio of the circular area to the disk diameter should be small. Thus, the design and fabrication of the disk gratings should be carefully done for accuracy. Of course, the precision of disk rotation is another important factor.

6.2.3 Blood Flow Velocity

One of the important applications of SFV in microscopic regions is measurements of blood flow velocity. In this application, red blood cells flowing in a vessel are imaged through a vessel wall onto a grating and their velocity is determined. Thus, the blood vessel being measured must be exposed to the SFV imaging system. Some examples of *in vitro* experiments have been reported for measurements of blood flow velocity and its spatial distribution in cylindrical glass tubes, by using a transmission grating [28, 30], a prism grating [40, 41], and the CCD line sensor or CMOS camera [62, 129]. Also, various *in vivo* measurements have been performed for arterioles of the foot web of frogs [24, 28], artery of the frog mesentery [37], arterioles and venules of the rabbit mesentery [39], arteriole in a rat cremaster muscle [29], and capillary at the human nail-fold (the skin overlapping the finger-nail at its base) [41, 62, 130]. For *in vivo* measurements, particular treatment is often required for the low signal-to-noise ratio due to the high concentration of blood cells and the temporal variation of velocity due to pulsations, physiological condition, and other unexpected motions such as postural fluctuations. To improve the quality of blood-cell images, an optical high-pass filter [30, 40], which eliminates dc light and enhances the edges of blood cells in the image, can be used in an SFV imaging system. The time-varying velocity of blood flows is effectively managed, for example, by using a narrow band-pass tracking filter [28] in the signal-analyzing system. The central frequency of the passband of this filter is voltage-controlled in the frequency tracking circuit so that it can follow the SFV signal frequency. Figures 6.6a and b [28] demonstrate the recorded variation and the corresponding frequency histograms, which were obtained by the transmission grating velocimeter of Fig. 6.1, of the blood flow velocity in a 30-μm diam arteriole when noradrenaline was put on the foot web. In result (**a**), the velocity clearly increases after an application of noradrenaline at time A and, then, begins to decrease at time B with the injection of water. Frequency histograms in result (**b**) were obtained at times (1)–(8) indicated in result (**a**). The central frequency of peaks of the histograms varies reasonably from (1) to (8) according to the temporal velocity variation.

6.2.4 Applications to Fluid Mechanics

In the original work on the SFV technique by Gaster [14], the application was already directed to free convection studies. By using the frequency-shifting technique with a rotating disk grating, the instrument developed in the work was capable of sensing the sign of flow and was used to measure the flow velocity distribution, including the reverse of flow, along the cross-sectional axis of a 2-inch diameter tube.

In some examples, specially devised spatial filtering probes have been developed to fit individual measuring circumstances. For measurements of an air–solid two-phase flow in a horizontal pipe, a fiber-array type spatial filtering probe [131] was constructed. The probe consists of 12 groups of graded-index quartz fibers having a core diameter of 80 μm which are aligned alternately for illumination and detection. The six groups of detecting fibers are differentially connected to an amplifier to eliminate the pedestal component. Although the number of grating lines results in $n = 3$ in this case, the probe was successfully used to obtain distributions of the solid particle velocity and concentration as contour lines over the cross section of the pipe. The particle concentration was determined by counting the number of particles passing through the probe volume in terms of a pre-determined intensity threshold for detected

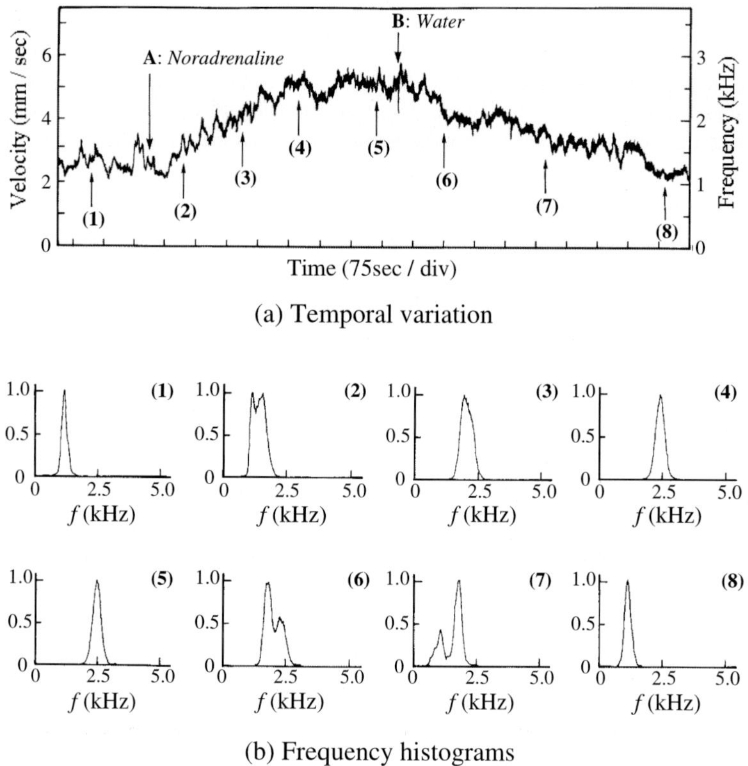

(a) Temporal variation

(b) Frequency histograms

Fig. 6.6. (a) Recorded variation and (b) the corresponding frequency histograms of the blood flow velocity in a 30-μm diam arteriole when noradrenaline was put on the foot web [28]

signals. A fiber probe is easy to vary in its measuring position owing to its flexibility and, thus, it is generally advantageous for large-scale or fixed flow channels.

A modified type of fiber-array spatial filtering probe [132] has been developed for measuring the local velocity of water flow in a horizontal pipe and channel. These probes, which are called hydrometric rod probe and plate probe in the literature, contain both a fiber illuminator and a fiber-array spatial filter and are used by insertion into a flow being measured. Thus, they are not nondisturbing devices and require calibration of the probe geometries in the velocity range of interest. The probe contains a measuring channel in itself, and end faces of the illuminator and the spatial filter are located flush with each of two opposite side walls of the channel. The fiber illuminator emits light via a fiber from an LED as a collimated beam with a graded-index lens toward the spatial filter through the channel. There is no lens used for imaging particles onto the spatial filter. Instead, the shadow of moving particles is projected onto the filter due to the collimated beam. The spatial filter consists of a differential arrangement of multimode fibers with a core diameter of $100\,\mu$m making the number of grating lines $n = 8$. Owing to the simple and robust mechanical construction of the probes, they were successfully used to measure the mean flow velocities in a 49-mm diameter horizontal pipe and in a horizontal channel with a cross section of $26 \times 56\,\text{mm}^2$ both $3\,\text{m}$ long.

Measurement of particle velocities in two-phase flows with high solids loadings requires a probe reaching different positions inside the flow. A spatial filtering probe using a CCD linear sensor [133], which is described in Sect. 5.7.2, has been constructed to measure the local particle velocity in a pilot-scale circulating fluidized bed riser with an inner diameter of $0.4\,\text{m}$ and total height of $15.6\,\text{m}$, which is used for studies in fluid dynamics. The solid particles being measured are quartz sand with a diameter range of 20–$500\,\mu$m and fluidized by air. The probe consists of a tube with an 8-mm outer diameter in which a small illuminating/detecting tip connected with light-guiding fibers, a receiving/focusing lens unit, and a CCD linear sensor are installed. By putting this probe into a lot of measuring holes at various heights of the riser alternately, local particle-velocity distributions in the bed were effectively obtained with a high measuring rate. Note that this probe also has the problem of disturbing the flow due to insertion.

The SFV technique by computer image processing has been applied to measurements of the oscillatory hydrodynamic flow in a nonlinear chemical reaction (Belousov-Zhabotinsky reaction) [125]. The technique employs a flexible change of pitch, shifting velocity and direction of the grating, and the compression/conversion and preprocessing of image data in real time and, thus, it successfully achieved increases in dynamic range, measurable time length, and measurement accuracy.

6.2.5 Flow Velocity Gradient

An interesting modification of transmission grating velocimetry has been made for measurements of the velocity gradient in a microscopic region [46]. Figure 6.7 shows schematically the principle of measuring a velocity gradient by using the SFV technique. The use of a modified spatial filter having two windows defines two probing cross-sectional areas in the object space. When flow velocities in the two areas are v_1 and v_2, the intensities of light passing through the two spatial-filter windows are periodically modulated with different frequencies f_1 and f_2, respectively, according to SFV principle. The photodetector output containing these periodic components is fed into an electric high-pass filter (HPF) to remove its low-frequency pedestal components. The filtered signal is then squared to produce four different periodic components with frequencies $2f_1$, $2f_2$, $f_1 + f_2$, and $f_1 - f_2$. By using an electric band-pass filter (BPF),the lowest frequency component $f_1 - f_2$ is selected and used to determine the velocity difference g between the two probe regions with the relation

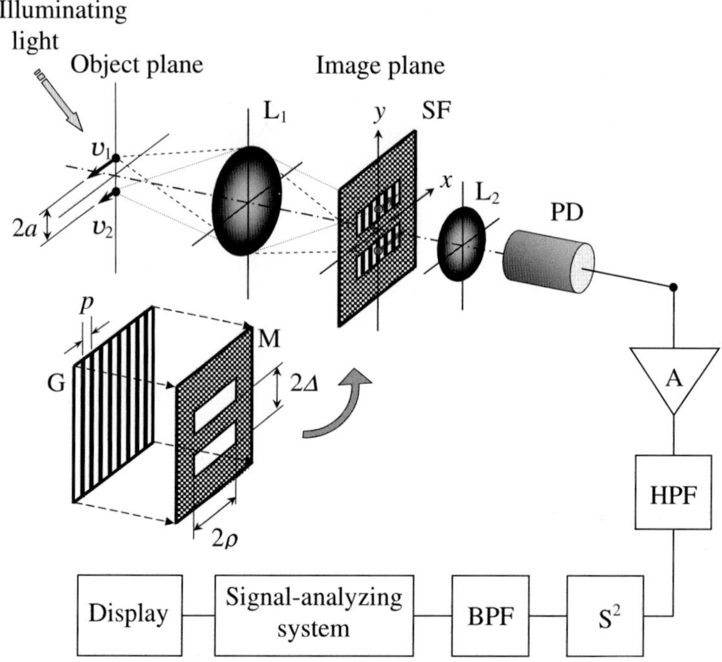

Fig. 6.7. Principle of measuring a velocity gradient using the SFV technique

$$g = \frac{v_1 - v_2}{2a} = \frac{p\,(f_1 - f_2)}{2aM}\,, \tag{6.2}$$

where $2a$ is the distance between the centers of the two probe areas, p is the grating pitch, and M is the imaging magnification. If the distance $2a$ is small compared to the whole flow area, the value g can be approximated as the velocity gradient.

Figure 6.8 [46] demonstrates measured distributions of the flow velocity and the velocity gradient of water in a 700-µm diameter cylindrical glass tube, obtained by using the transmission grating velocimeter shown in Fig. 6.1 where the modified spatial filter was exchanged with the original one. The area of two probed cross sections and their separation distance were set at $136 \times 27\,\mu\text{m}^2$ and $60\,\mu\text{m}$, respectively. Relatively good agreement was obtained between experiment and theory assuming Newtonian flow for both the velocity and the velocity gradient within errors of 2.1 and 11%, respectively. The errors are probably due to the range of velocities in the probe volume. Reducing the probe volume decreases the velocity range involved, but it may not guarantee the simultaneous passage of particles through the two probe volumes. There is a compromise between measurement accuracy and efficiency. The method has also been applied to measurements on the tip face of the 6-mm diameter rotating axis of a dc motor.

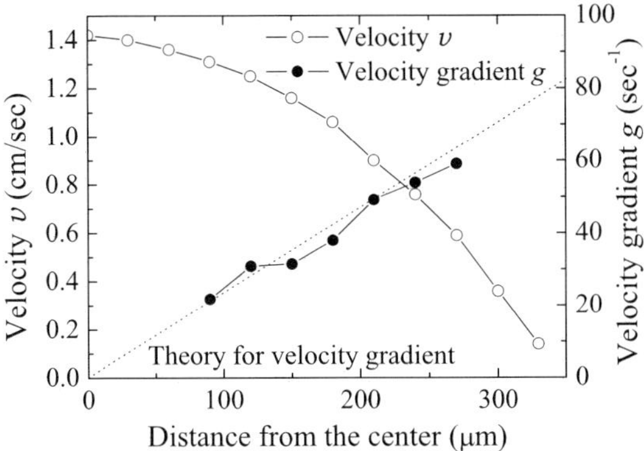

Fig. 6.8. Measured distributions of the flow velocity and velocity gradient of water in a 700-µm diameter cylindrical glass tube [46]

6.3 Measurements on Large Scales

The previous section describes SFV measurements in microscopic regions as typical and successful applications, in the same way as LDV measurements. Other examples of interesting SFV measurements are found in applications to large-scale regions such as rivers, debris, the ground, and so on. The LDV technique is generally unsuitable for velocity measurements in such wide regions due to the active sensing using a focused beam illumination, although some applications include airborne LDV and Doppler lidar. For these regions, passive sensing is more practical since it needs no illumination in large probing areas or beam scanning. This type of sensing can be realized with the SFV technique.

6.3.1 River Flows

Since the surface flow of water in a river can be imaged with a telescopic optical imaging system, the temporal intensity variation of the image is used for measuring the velocity of a river flow by using the SFV technique. Generally, the direction of a river flow is inconstant in time and space and, thus, the velocity should be measured in a vector. On the other hand, the flow velocity in a river yields a relatively low central frequency in the SFV system. By assuming a velocity of 10 m/s, imaging magnification of 0.002, and grating pitch of 0.25 mm, the central frequency is estimated at 80 Hz. Hence, a high-frequency response is not required in the SFV system. By taking account of these conditions, Itakura et al. [21] measured the flow velocity of a river surface by using a piled-type LC spatial filter, which is described in Sect. 5.5.2. To image a large area of the river surface onto the spatial filter on a bridge across the river, a Cassegrainian-type telescope having a focal length of 750 mm and a 250-mm diameter aperture was employed as the imaging system. The number of grating lines in the LC spatial filter was set at 5. Because moving LC cell arrays which were electrically driven produced undesirable shifting frequency components, they were filtered out in the signal processor. The magnitude and direction of the flow velocity were successfully measured, demonstrating the effectiveness of the SFV technique for measuring river flows.

The flow velocity under the water surface of a river was measured with the hydrometric plate probe [132] having a fiber-array spatial filter, which was described in Sect. 6.2.4. Before performing field measurements, the probe was calibrated with a test channel because it may disturb the flow being measured due to insertion. The results of the absolute profile of the flow velocity vector show the potential usefulness of the spatial filtering probe for measuring river flow dynamics.

6.3.2 Debris Flows

Debris flow is one of the natural random flowing phenomena and should quantitatively be observed for the purpose of disaster prevention. To perform the observation safely, the method employed should use remote sensing. Since the surface velocity of a debris flow can be unsteady in time, measurements must to be continuous. As one of the promising methods that meet these conditions, the SFV technique based on computer image processing has been studied [126–128], as illustrated in Fig. 6.9. A debris flow occurs unexpectedly and, thus, video images of the surface flow are recorded in the event. A series of frames are then analyzed by using appropriate processing software which operates on the basis of the SFV principle, as described in Sect. 5.8.2. Thus, field observation requires only the camera system and video recorder, and the image-processing site can be placed far from the observation site.

In the studies [126, 127], one velocity datum was extracted from the output signals of every 45 frames (i.e., every 1.5 s), by using the maximum entropy method (MEM). The processing software is able to optimize the parameters of the spatial filter: the grating pitch, the number of grating lines, and the window size to fit the intensity pattern of images being analyzed. The time-sequential velocity data were then recorded for 5 minutes. The result demonstrated the interesting finding that the debris flow velocity was initially low, was maximum approximately 1 minute after passage of the front, and then decreased gradually. A comparison study, also carried out between computer-based and hardware-based (such as a transmission grating) spatial filters, showed better accuracy with the computer-based type [128].

6.3.3 Aircraft

In 1963, Ator [12] presented the SFV technique for sensing the velocity of terrain images in the aerial photographic systems on aircraft. The velocity sensing of terrain images means determination of the flying speed of aircraft

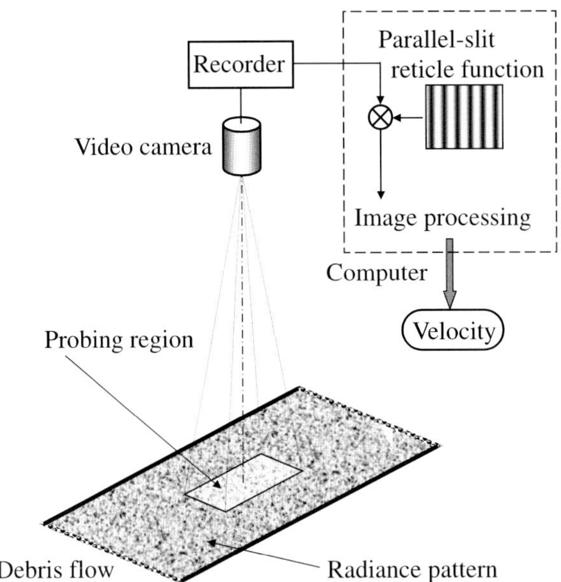

Fig. 6.9. Use of the SFV technique based on computer image processing for debris flows

with respect to the ground, assuming that the flight path of the aircraft is parallel to the terrain. Generally, the scattering of light by the ground produces a spatially random intensity distribution and, thus, the spatial frequency spectrum of the ground image can contain a range of frequencies which covers the central frequency of the spatial filter being used. Therefore, the terrain image can be passively sensed by the SFV technique without active illumination [13].

To cancel out the pedestal component that is caused by ambient terrain light conditions and atmospheric conditions, differential detection described in Sect. 5.1.2 is highly desirable for this application. Due to wind drift, the physical heading of an aircraft may not agree with the actual flight direction. To cope with this situation, two-dimensional velocity measurements should be performed, although directional discrimination may not be required because of unambiguity in the flying direction.

6.3.4 Vehicle

In the design and development of the automobile, precise measurements of the vehicle speed with respect to the ground are always indispensable. Normal speedometers of cars are available only for usual driving but are insufficient for various performance tests. In the antilock braking system, for example, the wheel slipping ratio which is given as a function of the wheel rotating speed and the ground speed of a vehicle is an important factor to be estimated

[134]. Hence, a noncontact direct sensor which can be carried on a vehicle is essential for determination of the ground speed. For these requirements, the SFV technique has been studied [134, 135] as a promising means. Optical images of road or ground surfaces have usually a random intensity distribution due to the surface irregularity whose spatial frequency spectrum extends over a certain range continuously from the dc component [135]. Thus, those surfaces can be sensed by the SFV technique under natural light.

There are specific treatments that are required for measurements of ground speed. Variations of the height of the sensor position on a vehicle may be a primary source of errors. To raise the robustness to height variation, a range of working distance and, thus, a large focusing depth are required in the optical imaging system, as discussed in Sect. 3.3. The use of a relatively large grating pitch is also effective for occasionally defocused images. Because a defocused image loses higher spatial frequency components but it may still contain lower components, these can be selected by a spatial filter having a low-frequency passband or a large grating pitch. A wide dynamic range of measurable speed is desired for practical uses such as nearly zero to a few hundred kilometers per hour. The temporal variation of the speed should also be recorded. To satisfy these requirements, real-time signal processing is employed such as a frequency counter together with a tracking filter or a frequency tracker, which are described in Chap. 4. The ambient light can be a source of low-contrast images and, thus, should be canceled out by using differential detection. The sensor is required to cope with various types of road surfaces, such as asphalt, concrete, sand, gravel, and snow. For this purpose, active illumination with light sources having a certain wavelength range such as LEDs is effective.

A prototype sensor [134] employed 22 LEDs of a 850-nm center wavelength for the illumination in a pulse mode driven at 60 kHz and synchronized detection for the purposes of enlarging the light power and reducing the effect of ambient light. In the sensor, the spatial filter was made with a slit array and a two-stage prism grating for differential detection. This sensor was then developed as a commercial product, which is called Ground View Sensor by its manufacturer [136], for recognition of road surfaces simultaneously with speed sensing [137, 138]. Figure 6.10 [137] depicts schematically the optical system of this sensor. Light rays from LED_1 and LED_2 are scattered and specularly reflected by the road surface, respectively, and the moving image of the surface is formed on the slit array. The scattered light is used for velocity measurements while the reflected light is used for recognition of the surface conditions among four different types: dry, wet, snow, and ice. The recognition is made by comparing the intensity of the pedestal component with that of the signal component. This sensor has achieved a velocity range of 0.1–120.0 km/h within an error of ±2% and a working distance of 300–350 mm. Another example uses halogen-lamp illumination and a spatial filter consisting of an integrated photodetector array [52], as shown in Fig. 6.11 [139].

Figure 6.12 [139] shows a photograph of the instrument setup on a car. Its specification discloses a velocity range of 1.5–320 km/h within an error

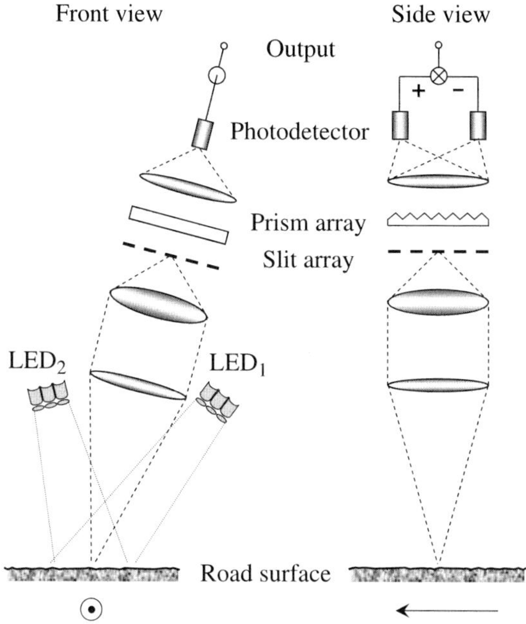

Fig. 6.10. Optical system of the Ground View Sensor [136] for recognition of road surfaces as well as velocity sensing [137]

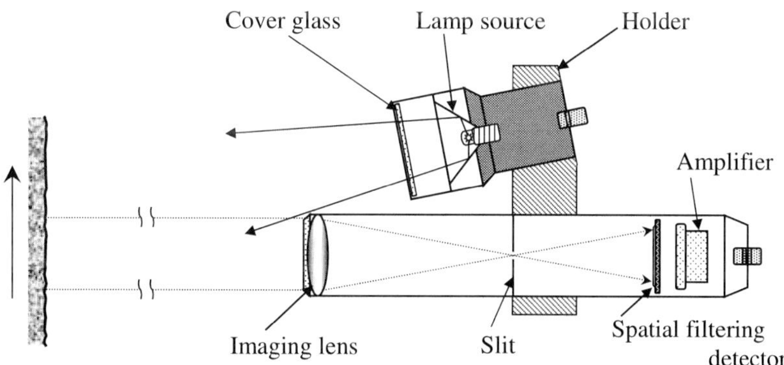

Fig. 6.11. Example of an instrument for measuring vehicle speed using halogen-lamp illumination and a spatial filter constructed with an integrated photodetector array [139]

of $\pm 0.5\,\%$, a working distance of $500 \pm 100\,\mathrm{mm}$ within an error of $\pm 0.1\,\%$, and a probe cross section of $46 \times 60\,\mathrm{mm}^2$. For instance, this instrument has been employed for the fine speed control of vehicles accompanying runners in marathon races.

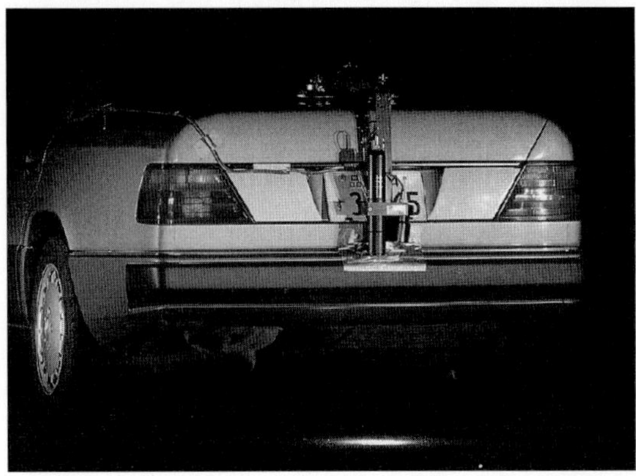

Fig. 6.12. Photograph of the instrument setup on a car [139]

6.3.5 Common Objects

Common moving objects including motorcars, bicycles, and pedestrians can also be sensed by the SFV technique [140]. However, there are some specific matters to be considered for this application. The moving speed of an object must be determined in only one pass through the detector's field of view. It is unavailable to improve the measurement accuracy in terms of any integration or averaging process for output signals. Common objects have a variety of shapes and sizes and, thus, the spatial frequency spectra of their image-intensity distributions are also different from each other. A careful design is required for the grating pitch so that its spatial frequency will appear within the frequency range of spectra of objects being measured. In images, a contrast of objects against a background is required to obtain a good SNR in output signals. The use of a polarizer can be effective in some cases. In the literature [140], a pedestrian, jogger, and cyclist 17 m distant and moving cars 52 m distant from the detector were probed for velocity determination within 4–5 % errors, by using a transmission grating, a photomultiplier, and a FFT-type spectrum analyzer. This application demonstrates the potential usefulness of the SFV technique for single-ended remote sensing and noncontact speed monitoring in production processes.

6.4 Potential Applications and Speckle Velocimetry

In principle, the SFV technique is able to measure the velocity of almost all moving objects that have nonspecular surfaces and can be optically or visually imaged with a certain contrast. Hence, there are various applications of the SFV technique, which have not yet been used as established means in science, engineering, and industry. This section briefly describes some objects which may be probed in SFV measurements and may be promising examples of the application.

6.4.1 Production Process

On production lines in manufacturing factories, a variety of materials, parts, and products run without intermittence. Precise control of the running speed of these objects is indispensable for efficient and scheduled production. Since some objects in a production process are easily imaged in an optical system or video system, their moving speeds can be measured with the SFV technique. Promising examples of such objects are iron plates, paper, cloth, thread, rubber, conveyor belts [20], V-belts [70], and so on. Figure 6.13 depicts typical output signals obtained from six different rough surfaces: ground glass, aluminum plate, wood, gum tape, white paper, and black paper, by using the transmission grating velocimeter shown in Fig. 6.1. Each piece of these materials was attached to a 120-mm disk rotating at constant velocity under the illumination of a He-Ne laser light and measured with the number of grating lines, $n = 10$ [28]. Burst-like periodic waveforms are successfully obtained from all the six surfaces in (**a**)–(**f**). In result (**f**), higher frequency noise is slightly larger than that in the others. This is due to the low contrast

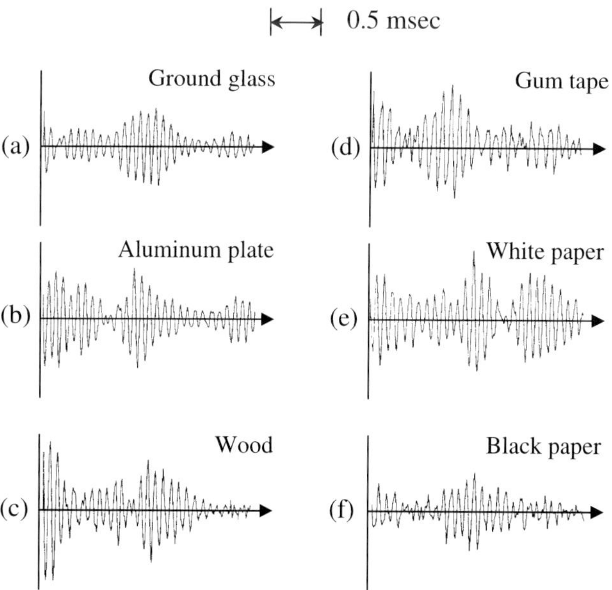

Fig. 6.13. Typical output signals obtained from six different rough surfaces; ground glass, aluminum plate, wood, gum tape, white paper, and black paper, by using a transmission grating velocimeter

of the image of the black paper. Therefore, the applicability depends on the quality of the images. High-contrast images produce better SFV signals. As discussed in Sect. 2.7 including Fig. 2.25, this means that the intensity distribution pattern of images should contain a sufficient magnitude of spatial frequency components around the fundamental frequency $\mu_0 = 1/p$ of the spatial filter being used. To satisfy this condition, the choice of the illuminating direction, light source, imaging magnification, and grating pitch may be effective very often.

Some machines used in factories are in translational and rotational motion which can also be probed by the SFV technique. The rotating velocity of a motor axis has been measured at its tip face by using a transmission grating velocimeter [28]. For measurements of rotating velocity, a radial-type spatial filter as shown in Fig. 5.37 [52,107], is quite effective since the angular velocity in rpm is obtained directly from the central frequency of the output signals independently of the imaging magnification. Misalignment between the two center axes of a radial spatial filter and a projected rotating image reduces the amplitude of the periodic signal component having a desired frequency, and it produces subsidiary frequency components which are undesired. However, this is used for detecting the center position of a rotating axis.

6.4.2 Rain and Snow

As routine atmospheric phenomena, rain and snow are observed in natural scenes because raindrops and snowflakes scatter light. Optical images of the drops and flakes are sometimes of poor quality due to defocusing and low contrast. Unless axial resolution is required in an object space, a large focusing depth is useful for moderating the defocusing effect. Since the light intensity scattered by raindrops or snowflakes is weak, their images generally tend to be obscured in a background scene. To obtain high-contrast images, active illumination is necessary. When those scatterers are clearly imaged in the background scene, their falling velocity may be measured by the SFV technique. Thus, imaging is the key for this application.

As an object similar to the above, smoke ascending to the sky may also be probed by the SFV technique. For example, smoke from a chimney stack of a factory can be imaged without difficulty. Then the random intensity fluctuation of smoke images is used for spatial filtering. Other potential objects similar to those in the above application include dust, ashes, fire, spray, and so on. For all these measurements, the key to success is imaging with the better contrast against the background.

6.4.3 Micromachines and Biological Samples

In the technology of micromachines, motion must be precisely controlled on a microscopic scale. Speed monitoring should be done in a nondisturbing manner since such small machines are quite easily influenced by external forces and momentum. The SFV technique is a promising tool for monitoring the moving velocity if the motion of micromachines can be imaged with the necessary magnification. For example, the rotating velocity of micromotors and gears or the velocity of micropumps may be determined by SFV measurements.

It is known that small dielectric particles can be manipulated by the radiation pressure of laser light. In an optical manipulation experiment, a highly focused laser beam is used to illuminate a particle for the manipulation and, thus, the particle scatters the light beam with a rather high intensity. Using a high-resolution optical microscope to receive this scattered light, the particle motion can be viewed while being manipulated, without the necessity of additional illumination. By applying the SFV technique to the images of moving particles obtained by a microscope, the velocities of the particle motion, including translation and rotation, can be measured. Micron-sized biological samples such as bacteria operated by radiation pressure are also interesting objects to which the SFV technique is applied in the same way. Note in this kind of application that the laser light used for manipulating particles works simultaneously as the illuminating light for imaging in SFV measurements. If the LDV technique is applied to measurements of particle motion, the radiation pressure is affected by the focused laser beam used in the LDV optical system.

Recently, biological molecular motors are receiving much attention from researchers in various fields. In studies of molecular motors, a quantitative estimation of linear or rotational motion is necessary for the analysis of their moving mechanism and energy metabolism. This experiment requires a microscopic observation system with high resolution and high magnification. To keep biological samples from being optically damaged, the wavelength and the intensity of the illuminating light should be carefully chosen. Thus, the imaging of samples with necessary contrast and resolution promotes an application of the SFV technique to motion measurements of molecular motors.

6.4.4 Laser Speckle Velocimeter

As described in Sect. 2.7.3, the SFV technique is used to determine the moving velocity of a translational speckle pattern. With the known relation between the velocities of a moving object and the corresponding moving speckle patterns, the spatial-filtering-type laser speckle velocimeter can be constructed. Stavis [63] first demonstrated the velocity determination of a moving diffuse object through SFV measurements in which a simple transmission grating was applied to translational speckles formed in the diffraction field. Komatsu et al. [141] represented the velocity measurements using dynamic speckles produced in the image plane when the transmission grating was placed in the diffraction field in front of the imaging lens. Instead of the grating, a linear photodiode array [111, 112] was also used to simplify the system and to suppress the pedestal component by differential detection.

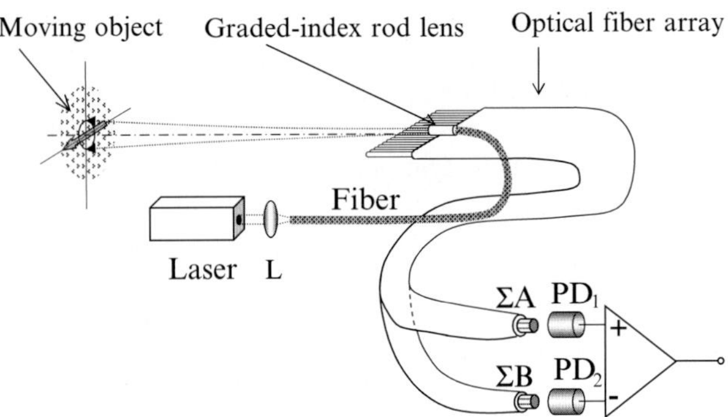

Fig. 6.14. Use of the optical-fiber-array spatial filter for a fiber-type laser speckle velocimeter

Using the optical-fiber-array spatial filter, which was described in Sect. 5.4, Hayashi and Kitagawa [142] developed a fiber-type laser speckle velocimeter for moving diffuse objects, as shown in Fig. 6.14. The velocimeter employs a graded-index fiber having an 80-µm core diameter for the illumination and the spatial filter made with 20 step-index fibers of 90-µm core diameter each. Translational speckles produced by the object are detected in the diffraction field with the fiber array and processed by the differential connection. To employ the specific parameters for the illuminating and observing geometry, velocity measurements with this velocimeter are insensitive to fluctuations of the distance between the object and the velocimeter. Usually, the moving velocity of a translational speckle pattern depends on the illuminating and observing geometry including the distance between the object and the velocimeter [143]. By detecting the speckle motion at two different positions with two fiber-array spatial filters, the object velocity and distance can be determined simultaneously [144]. Instead of using two spatial filters, a set of two illuminating fibers giving different beam propagations and one fiber-array spatial filter is also available for the same purpose [145].

Recently, Jakobsen and Hanson [146] developed a lenticular-type laser speckle velocimeter for measurements of in-plane translation or rotation of solid structures. This velocimeter detects translational speckles in the image plane by using a lenticular grating followed by a spherical lens and two photo detectors for pedestal removal. The detailed design of an optical system including the lenticular-type spatial filter is useful for implementation of a practial velocity sensor. The velocimeter was then elegantly miniaturized [147] by the combination of a vertical cavity surface emitting laser (VCSEL) and a microlenticular grating with a pitch of 15 µm. In this system, translational speckles are detected in the diffraction plane, and no imaging optics is needed. The well-designed optical configuration realizes an almost working-distance-invariant measurement. The system is low cost and robust, and, thus, it seems to be advantageous for various industrial applications.

6.5 Derivative Measurements

Though the main theme of this book is velocity measurements by the spatial filtering technique, the SFV technique is applied to measurements of some other physical values that are important in science and engineering. This section simply reviews particle sizing, focus detection, distance measurement, and speckle displacement sensing, which are done derivatively or through the spatial filtering detection of motion.

6.5.1 Particle Sizing

As described in Sect. 2.6, the visibility of output signals decreases as the ratio of the particle diameter to the grating pitch increases. This visibility change follows some different mathematical functions in accordance with the particle shape and the grating transmittance function, as shown in Fig. 2.23. In any case except for the Gaussian-shape image, there exist specific values of the diameter-to-pitch ratio with which the visibility takes the first zero or minimum. Wang and Tichenor [49] applied this property to particle size measurements using a variable-pitch grating. For a circular image with the diameter $2b$ passing on the sinusoidal-transmittance grating, the visibility is given by (2.56). The first null in J_1 or the Bessel function of the first order occurs where the argument of J_1 equals 3.83. Then, the grating pitch p_0 that makes the first-null visibility for the given diameter $2b$ is simply expressed by

$$p_0 = \frac{2\pi b}{3.83} = \frac{2b}{1.22} \ .$$

(6.3)

Figure 6.15 illustrates the passing of a circular image on a varying-pitch grating [49]. Figure 6.16 shows a model of the signal waveform obtained with such a grating in which a point of the first null is identified [49]. With the knowledge of local grating pitches, the image diameter is determined by counting the number of grid cycles before the first null in visibility and identifying

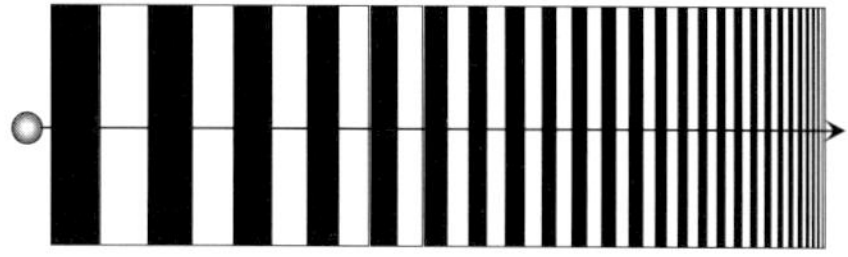

Fig. 6.15. Passing of a circular image on a varying-pitch grating

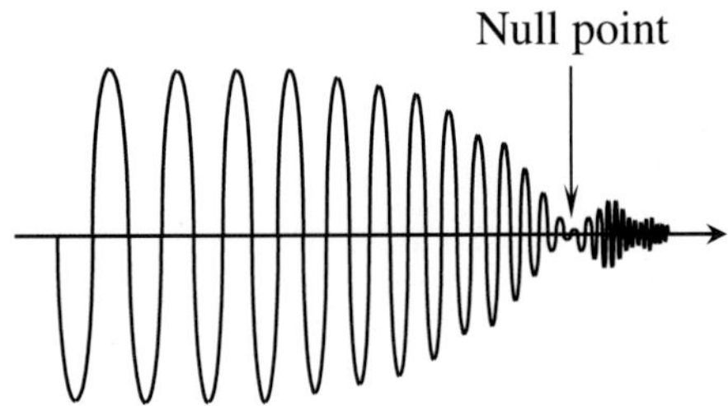

Fig. 6.16. Signal waveform simulated for the situation of Fig. 6.15

the grating pitch p_0. In actual measurements, a point with null visibility is often unclearly observed in signals due to noise. Therefore, a high SNR in signal recording and statistical estimation in data processing are required in applications. It seems rather difficult to fabricate a varying-pitch grating with the required precision and chirping rate. Michel [62] realized electronically a grating of this type by using a CMOS spatial filter in which the varying pitch is flexibly arranged by gradually changing the number of pixel summations in data processing.

There is another approach with the normal transmission grating having a constant pitch. The visibility curves in Fig. 2.23 contain certain ranges of the diameter-to-pitch ratio where the visibility monotonically decreases as the ratio increases. In these ranges, measurements of the signal visibility provide a means for particle size determination using a known grating pitch [148], simultaneously with velocity sensing. This method has already been introduced in LDV for the same purpose [149,150]. Using (2.56), for example, a measured value for the visibility specifies the corresponding value for the argument x of the Besinc function, $J_1(x)/x$. Then, the particle-image diameter is determined by the relation of $2\pi b/p = x$. Unfortunately, the measurable range of particle size with a given grating pitch is rather limited because the diameter-to-pitch ratio is a multivalued function of visibility over much of the range, as shown in Fig. 2.23. Particle sizing with the SFV technique assumes the existence of a single particle in the probe volume. Measurements of many particles may be possible in some limited cases [148].

6.5.2 Focus Detection

In optical imaging systems, a focusing error increases the size of a particle image and decreases its clearness or sharpness due to blurring of the edge. This defocusing effect reduces higher frequency components of the image intensity distribution in the spatial frequency domain. Then, the effect degrades visibility since the relative intensity at the central frequency of the spatial filter decreases with respect to that of the pedestal component. On the basis of this nature in the SFV technique, the focus position can be detected

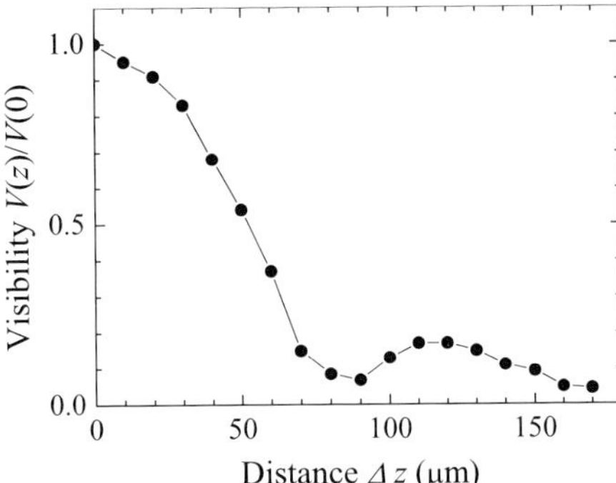

Fig. 6.17. Measured example of visibility as a function of the plate position shown by the distance Δz from the focus plane (from [148], Masson 1990)

by measuring visibility. Tsutsumi [16] found theoretically and experimentally that searching the maximum average power of the periodic signal component gives the focusing position for infrared aerial camera systems. The theoretical investigation [50] was made to derive the optimum spatial filtering characteristics and to propose the binary checkerboard reticle as a realized example. Some experiments [23, 148] were performed to describe the effect of focusing errors on visibility using a glass plate having small particles on the surface in a microscopic region. Figure 6.17 [148] shows the measured visibility plotted as a function of the plate position with the distance Δz from the focus plane. With an increasing distance Δz or focusing error, visibility varies and seems to follow a Besinc function. This result indicates that monitoring the maximum visibility gives the focus position of an object being measured at the same time as the velocity determination.

6.5.3 Distance Measurement

Once the focus position is detected by the above-mentioned method, the distance from the principal plane of an imaging lens to the plane of a moving object can be determined by the simple imaging equation (3.3), in which the distance from the principal plane to the image plane and the focal length of the lens are assumed known. A more realistic idea [51, 52] is to use a two-channel imaging system with different magnifications, as depicted in Fig. 6.18. A half mirror (HM) in an optical imaging system forms images of the moving object onto the planes of transmission gratings G_1 and G_2 with magnifications $M_1 = b_1/(d + a_1)$ and $M_2 = b_2/(d + a_2)$, respectively, in terms of notations in the figure. When the grating pitches for G_1 and G_2 are p_1 and p_2, respectively, and f_1 and f_2 are the central frequencies of detector outputs in channels 1 and 2, respectively, the velocity v of the object is expressed as

$$v = \frac{(d + a_1)\, p_1 f_1}{b_1} = \frac{(d + a_2)\, p_2 f_2}{b_2} . \tag{6.4}$$

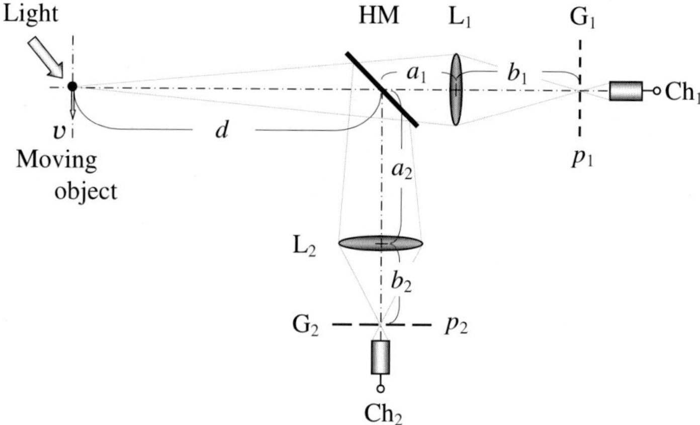

Fig. 6.18. Use of a two-channel imaging system with different magnifications for distance measurements

From this relation, the velocity v and the distance d from the position of the half mirror to the object are derived as

$$v = \frac{(a_1 - a_2)\, p_1 p_2 f_1 f_2}{b_1 p_2 f_2 - b_2 p_1 f_1} \,, \tag{6.5}$$

$$d = \frac{a_1 b_2 p_1 f_1 - a_2 b_1 p_2 f_2}{b_1 p_2 f_2 - b_2 p_1 f_1} \,. \tag{6.6}$$

A simple and practical example is the case of $p_1 = p_2 = p$ and $b_1 = b_2 = b$, which gives equations

$$v = \frac{(a_1 - a_2)\, p f_1 f_2}{b\,(f_2 - f_1)} \,, \tag{6.7}$$

$$d = \frac{a_1 f_1 - a_2 f_2}{f_2 - f_1} \,. \tag{6.8}$$

Since the distances a_1, a_2, and b are usually known, measurements of the two frequencies f_1 and f_2 provide a simultaneous determination of the velocity and distance. This method is effective for SFV measurements in remote sensing such as flying objects and atmospheric observations.

Instead of using an imaging system, this idea can be realized with speckle translations in two different diffraction planes [151]. The velocity of speckle motion depends on the distance between a moving object and an observation plane. By placing the two transmission gratings G_1 and G_2 in diffraction planes with different distances from the object, two different-frequency SFV outputs are obtained, which are used to measure the distance to the object. The optical system configuration makes measurements independent of the object's velocity.

6.5.4 Displacement Sensing by Speckle

Spatial filtering detection of translational speckles can be used for velocity measurements, as described in Sect. 6.4.4, and also for other dynamics including displacement, deformation, and vibration. When the diffuse object under the illumination of laser light is subjected to displacement, the speckle pattern produced by the object is also displaced. Since the displacement is a position change due to translational motion for a finite time, the definite integral of the motion velocity determined by SFV measurements gives the displacement. In practice, however, it is more convenient to count the number k of cycles in SFV periodic signals after pedestal removal during the displacement. With the known grating pitch p and imaging magnification M, the displacement Δs is simply obtained as

$$\Delta s = \frac{kp}{M} \, , \tag{6.9}$$

regardless of the object's velocity in the displacement. Yamaguchi et al. [152,153] used this method for their laser speckle strain gauge, in which a photodiode array was used as the spatial filter to realize electronically differential detection and frequency shifting. They applied the gauge to strain measurements of high-polymer films in various directions under loading and evaluated their Poisson ratios. This type of displacement sensing can also be applied to any movement or displacement of linear interference fringes. Then, the SFV technique is potentially useful for various interferometric measurements such as length, vibration, temperature [154], and even in Fourier transform spectroscopy. Jakobsen et al. [155] applied their lenticular-type spatial filtering velocimeter to submicron in-plane vibration measurements. In their system, two differential output signals with a mutual phase shift of $\pi/3$ provide the phase information, which is used for monitoring real-time vibration.

6.6 Related Techniques

This section is devoted to describing some similar velocimetric techniques that are not usually categorized in spatial filtering velocimetry. There are possible methods which can make the spatial filtering effect to produce the same periodic signal as obtained in the SFV technique. They employ a grating-like illumination, Young's fringes, and laser diode array, instead of a spatial filter. Mention should also be made of a modified type of spatial filtering effect which detects the entire spectral width of frequency components generated by random motion.

6.6.1 Grating Illumination

When a particle image with constant velocity crosses the lines of a transmission grating, the intensity of the transmitted light is periodically modulated to yield SFV output signals. The same mechanism as this intensity modulation in the image plane is realized in the object plane where a particle with constant velocity crosses the lines of some types of fringes, but not "Doppler" interference fringes. Such fringes can be formed in white light by projecting the image of a transmission grating into the probe volume. Thus, the illumination consists of alternate bright and dark bands. The light scattered by moving particles shows a periodic intensity modulation and is simply received by a photodetector without the grating in front, yielding quite the same type of output signals as in the SFV technique. This method, called the "fringe image technique," was first proposed and studied by Ballik and Chan [43]. The scheme of fringes in the probe volume is quite similar to the real-fringe interpretation of the LDV or the differential LDV technique. Then, studies [5, 43–45] were mainly comparison with LDV. Geometric light scattering by particles under grating illumination was theoretically and experimentally investigated [156–158], aiming at velocity and/or size measurements of bubbles and drops in dispersed two-phase flows.

The fringe image technique does not require, in principle, imaging of particles in front of the photodetector, and the scattered light may straightforwardly be received in any direction. The disadvantages of using grating illumination are that it is difficult to define clearly a depth of the probe volume and that the line spacing is not consistent over the entire probe volume due to the imaging properties of grating lines. The fringe image technique is a type of "active sensing" which means that measurements are made by illuminating a target with light. Thus, this technique is unavailable for extremely small particles in a microscopic region and natural objects on large scales such as river and ground, since it is practically difficult or impossible to employ grating illumination.

Chan and Ballik [159] also studied an application of Fourier images to the formation of real fringes in a probe volume. When a periodic object (e.g. Ronchi grating) is illuminated by a collimated monochromatic light beam, Fourier images with spatial periodicity are obtained as a result of Fresnel diffraction from the periodic object. Thus, the grating-like illumination is realized in the probe volume and used for velocity measurements by the principle of the fringe image technique. Note that Fresnel diffraction gradually changes to Fraunhofer diffraction as the object-to-image distance becomes sufficiently large. The major property of the Fourier image is its wide range in both depth and cross-sectional area because of Fresnel diffraction. An equivalent illumination method was used for measuring the velocities of tennis balls and cars by different researchers [160].

6.6.2 Double-Exposure Specklegram

There is an alternative way to produce fringes in the probe volume. Kadono et al. [161, 162] proposed a velocimeter using a double-exposure specklegram.

In speckle photography or single beam speckle interferometry [163], a double-exposure specklegram is made by taking a doubly exposed photograph of two speckle patterns, one the original and another obtained by shifting the position of a recording film with a certain distance. Then, the specklegram has a number of pairs of two identical speckle grains which act as a double-pinhole pattern. By illuminating the specklegram with laser light, the diffraction pattern of the specklegram obtained is a form of Young's fringes in the probe volume. The intensity of light scattered by particles moving across the fringes is sinusoidally modulated in the same way as using grating illumination. Receiving the scattered light produces output signals whose central frequency is proportional to the moving velocity of the particles.

The spacing of Young's fringes is flexibly varied by changing the position of a specklegram placed in the illuminating optical system. It is also easy to make double-exposure specklegrams in laboratories. A problem with this method is the speckle noise that is superposed onto Young's fringes. Speckle grains in the pattern of Young's fringes result directly in noise in the output signals which degrade the quality of sinusoidal waveforms and broaden the frequency spectrum. To reduce the effect of speckle noise, the fringe spacing should be sufficiently made larger than the average size of the speckle grains.

6.6.3 Diode Array Velocimetry

A different and more direct way to form a spatially periodic illumination pattern is the use of a laser diode array [164], which generates multiple (four or more) discrete laser spots separated by short distances (20–100 μm) in the probe volume. The passing of particles through the multiple spots produces intensity-modulated periodic signals in the photodetector output. The central frequency of the output signals and the separation of laser spots in the probe volume determine the particle velocity. This technique, discussed under comparison with LDV in aiming at flow velocity measurements, is called diode array velocimetry (DAV) by the authors. Two basic types of laser diode arrays are used for this technique: phase-coupled stripe laser diode and individually addressable laser diode arrays. Although both types have advantages and disadvantages, the latter type is desirable for applications due to electric controllability of each of the outputs in the multiple laser spots.

In comparison with grating illumination or the specklegram, the laser diode array is attractive in illuminating light power. The Gaussian intensity distribution of each spot in the probe volume is also advantageous since theoretically it causes no high-frequency harmonics in output signals. A drawback of this technique may be difficulty in defining the probe volume. All spots should be formed in a line in the moving direction of particles. Thus a focusing lens with high quality is required. A probe volume depth is determined by the focusing depth of each beam spot and, thus, its clear definition seems difficult.

6.6.4 Random Pattern Velocimetry

There are modified types of the spatial filter which do not generate a periodic signal oscillating at a specified central frequency but a random signal having a range of frequencies from the dc to the cutoff region. In this case, the range of frequencies is related to an overall estimation of various velocity components or random motion. Three types of filters have been proposed and studied for sensing random motions: two-dimensional M-sequence random pattern [107], randomly distributed multiaperture pattern [165], and photographic speckle pattern [165].

M-sequence functions can be used for generating pseudorandom variables [166, 167]. Using M-sequences, a two-dimensional ternary random pattern weighted with three levels of -1, 0, and $+1$ has been proposed and realized by a two-dimensional silicon photodiode array as a spatial filtering detector. Pixels of the detector array are electrically connected to an electronic circuit via aluminum electrodes and load resistors so that the array will contain a spatial weight pattern which is predetermined according to a two-dimensional ternary M-sequence. When the random motion of small objects such as particles or bacteria is imaged onto the 2-D random pattern detector, randomly modulated intensity signals are generated by assuming that the motions of objects are statistically independent of each other. As the moving velocity of objects increases, frequency components of this modulation become higher. Thus, the range or spread width of a power spectrum gives a "mean velocity" or activity of the random motion.

Similar random signals can also be produced by detecting the image of random motion through a random pattern which has spatially random transmittance. Two examples of the random pattern were studied: a pattern of randomly distributed multiple apertures and photographic film of a speckle pattern. Although the above-mentioned three kinds of random patterns are expressed by functions mathematically different from each other, the principle of estimating the random motion is the same. Instead of the power spectrum, the autocorrelation function of output signals is also analyzed for the estimation. To estimate the mean velocity, the average frequency or the cutoff frequency of the power spectrum and the correlation time of the autocorrelation function are used. This type of velocimetry is useful for overall or macroscopic measurements of random movements such as the motion of small animals and micro organisms such as bacteria, the Brownian motion of small particles in a liquid, and pedestrian motion on a road.

6.7 Brief Comparison with Laser Doppler Velocimetry

Among optical velocimetric techniques, the SFV technique closely compares with the differential-type LDV technique due to their similar methods of measurement. The differential-type LDV system is very often interpreted by the real fringe model [5], in which the measuring principle seems to be almost the same as that of the SFV technique. However, note that SFV uses the spatial filtering effect of moving images whereas LDV is based on the Doppler effect on laser light scattered by moving objects. The basic operations, for producing signals, are image detection with a transmission grating in SFV and interference with laser light under the coherence condition [5] in LDV. They have different results in some situations. For example, as the number of scattering particles approaches infinity such as in a uniform cloud of particles, SFV produces almost no periodic variation in the intensity of the detected light but LDV keeps some residue of signals due to the interference of light fields scattered from different particles. LDV periodic signals are sinusoidal, but SFV signals sometimes contain higher harmonics due to the binary transmittance of a grating.

Though LDV requires a coherent light source due to the principle, SFV holds a choice of coherent and incoherent light sources. Focused laser beams are advantageous for high-intensity illumination but unavailable for large-angle illumination. For the latter purpose, conventional incoherent light sources are rather convenient. For SFV, it is possible to select an appropriate light source in accordance with measurement circumstances. Wavelength fluctuations in the light source are an essential problem with the LDV but are insignificant for SFV. Because there is no requirement of coherency for the source, SFV works in a passive mode or a condition with no active illumination. If a moving object is imaged with a reasonable contrast under natural light, sunlight, or ambient light, SFV signals are obtained and the velocity is determined. This is quite different from LDV and makes SFV available for a wide range of applications.

It is well known that high spatial resolution is one of the useful advantages of LDV. By using a microscopic imaging system together with a well-designed focusing depth and a spatial-filter window, SFV may approach the same order of resolution as LDV. Conversely, for a large probe-sectional area, SFV is easily available with a telescopic imaging system but LDV has difficulty such as in high-speed two-dimensional scanning. Theoretical and experimental studies show that SFV has measuring accuracy comparable with LDV in general cases. The choice between SFV and LDV for accuracy should be determined by the objects being measured and the measurement conditions. The quality of imaging is the primary factor for deciding the accuracy of SFV measurements. For signal processing, there is little difference between SFV and LDV techniques. In fact, most LDV signal processors can be used straightforwardly for SFV. However, a wrong choice of processing method causes substantial errors in the results. Note that different types of signals require different signal-processing methods.

The most practical advantages of the SFV technique are that the optical system is simple and mechanically stable and, thus, robust and easy to operate. The instrumental system is usually inexpensive, and commercialization is expected without high cost. Although LDV is widely known as the most

Table 6.1. Comparison of SFV and LDV techniques

	SFV	LDV (real-fringe type)
Principle	Spatial filtering	Doppler effect
Light source	Incoherent or coherent	Coherent only
Optical system	Imaging	Interferometry
Requirement	Focusing	Coherence condition
Basic operation	Image detection	Interference detection
Sensing scheme	Passive or active	Active only
Advantages	Simple and easy, inexpensive, wide-area probing	High spatial resolution
Disadvantages	Higher harmonics contained	Complicated alignment, skill required

representative optical velocimetric technique, its optical system is relatively complicated, costly, and sensitive to mechanical disturbance, and the optical alignment and measuring operation require skill and experience. Many measurements that are conventionally made by LDV may be favorably and more efficiently done by SFV. There are also many situations where SFV shows better performance or LDV is unavailable. Therefore, spatial filtering velocimetry has enormous potentialities for both laboratory and industry applications in various fields of science and engineering. Table 6.1 briefly compares the SFV and LDV techniques in some fundamental items.

A

Fourier Analysis

Fourier analysis is a useful and universal mathematical tool for studies of linear phenomena and is widely used in science and engineering. This tool is able to decompose a complicated system input into elementary inputs and to analyze the total response of the system in terms of a linear combination of the individual known responses to those inputs.

A.1 Fourier Series

A periodic function $f(t)$ of a period T and a fundamental frequency $u_0 (= 1/T)$ is expressed as

$$f(t) = \frac{a_0}{2} + \sum_{n=1}^{\infty} \left[a_n \cos\left(2\pi n u_0 t\right) + b_n \sin\left(2\pi n u_0 t\right) \right] , \qquad \text{(A.1)}$$

where t is one independent variable and n is an integer, and

$$a_n = \frac{2}{T} \int_{-T/2}^{T/2} f(t) \cos\left(2\pi n u_0 t\right) \mathrm{d}t , \quad n = 0, 1, 2, \ldots , \qquad \text{(A.2)}$$

$$b_n = \frac{2}{T} \int_{-T/2}^{T/2} f(t) \sin\left(2\pi n u_0 t\right) \mathrm{d}t , \quad n = 1, 2, 3, \ldots . \qquad \text{(A.3)}$$

In the expression using a complex-valued function, the above expansion is written as

$$f(t) = \sum_{n=-\infty}^{\infty} c_n \exp\left(\mathrm{i}2\pi n u_0 t\right) , \qquad \text{(A.4)}$$

$$c_0 = \frac{a_0}{2} , \qquad \text{(A.5)}$$

$$c_n = \frac{1}{T} \int_{-T/2}^{T/2} f(t) \exp\left(-\mathrm{i}2\pi n u_0 t\right) \mathrm{d}t . \qquad \text{(A.6)}$$

An example of the rectangular function defined by

$$f\left(t\right) = \begin{cases} A, & \left(n - \dfrac{1}{4}\right) T \le t \le \left(n + \dfrac{1}{4}\right) T, \\ 0, & \text{otherwise} \end{cases} \tag{A.7}$$

is given by

$$f\left(t\right) = \frac{A}{2} + \frac{2A}{\pi} \\ \cdot \left[\cos\left(2\pi u_0 t\right) - \frac{1}{3}\cos\left(2\pi 3 u_0 t\right) + \frac{1}{5}\cos\left(2\pi 5 u_0 t\right) - \cdots\right]. \tag{A.8}$$

A.2 Fourier Transform

The Fourier series expansion of a periodic function can be extended to a non-periodic function with the operation of the period $T \to \infty$. Here, c_n given by (A.6) is expressed as

$$c_n = \frac{1}{T} F\left(u_n\right), \tag{A.9}$$

where $u_n = n u_0$ and

$$F\left(u_n\right) = \int_{-T/2}^{T/2} f\left(t\right) \exp\left(-i 2\pi u_n t\right) dt. \tag{A.10}$$

Then, (A.4) is rewritten as

$$f\left(t\right) = \sum_{n=-\infty}^{\infty} \frac{1}{T} F\left(u_n\right) \exp\left(i 2\pi u_n t\right). \tag{A.11}$$

By making $T \to \infty$ or $u_0 = 1/T \to 0$, the discrete variable $u_n = n u_0$ turns out to be a continuous frequency u. Thus, (A.10) is converted to

$$\lim_{T \to \infty} F\left(u_n\right) = F\left(u\right) = \int_{-\infty}^{\infty} f\left(t\right) \exp\left(-i 2\pi u t\right) dt, \tag{A.12}$$

where $F(u)$ is generally called a frequency (or Fourier) spectrum. By taking also account of $\sum \to \int$ and $1/T = u_0 = u/n \to du$ with the operation of $T \to \infty$, (A.11) is written as

$$f\left(t\right) = \int_{-\infty}^{\infty} F\left(u\right) \exp\left(i 2\pi u t\right) du. \tag{A.13}$$

Equations (A.12) and (A.13) indicate that the spectrum $F(u)$ is the Fourier transform of the signal $f(t)$ and that $f(t)$ is the inverse Fourier transform of $F(u)$. It is usually said that the two functions $f(t)$ and $F(u)$ have the relation of a Fourier transform pair.

A.3 Two-Dimensional Expression

In various optical phenomena, the propagation of light is naturally treated by light distributions in space, especially in a plane normal to the optical axis. Then, Fourier analysis is applied to functions of two independent variables in optics. The Fourier transform of a complex two-dimensional function $f(x, y)$ is defined by

$$F(\mu, \nu) = \iint_{-\infty}^{\infty} f(x, y) \exp\left[-i2\pi(\mu x + \nu y)\right] dx dy , \qquad (A.14)$$

where μ and ν denote frequencies in the space domain (x, y), which are called spatial frequencies, and $F(\mu, \nu)$ is a frequency spectrum in two dimensions. The inverse Fourier transform of the function $F(\mu, \nu)$ is given by

$$f(x, y) = \iint_{-\infty}^{\infty} F(\mu, \nu) \exp\left[i2\pi(\mu x + \nu y)\right] d\mu d\nu . \qquad (A.15)$$

For a brief discussion of the existence conditions in the above two definitions, see the book by Goodman [55].

A.4 Fourier Transform Theorems

The definition (A.14) of the Fourier transform gives some useful mathematical theorems. In the following, the Fourier transform and the inverse Fourier transform are represented by notations $\mathscr{F}[\cdots]$ and $\mathscr{F}^{-1}[\cdots]$, respectively. Also, the Fourier spectra of two-dimensional functions $f(x, y)$ and $h(x, y)$ are expressed by $F(\mu, \nu)$ and $H(\mu, \nu)$, respectively.

1. *Fourier integral theorem*

$$\mathscr{F}\mathscr{F}^{-1}[f(x, y)] = \mathscr{F}^{-1}\mathscr{F}[f(x, y)] = f(x, y) . \qquad (A.16)$$

2. *Linearity theorem*

$$\mathscr{F}[af(x, y) + bh(x, y)] = a\mathscr{F}[f(x, y)] + b\mathscr{F}[h(x, y)] . \qquad (A.17)$$

   The linear combination of functions holds in the transform.
3. *Similarity theorem*

$$\mathscr{F}[f(ax, by)] = \frac{1}{|ab|} F\left(\frac{\mu}{a}, \frac{\nu}{b}\right) . \qquad (A.18)$$

   A stretching of the space coordinates (x, y) leads to a contraction of the frequency coordinates (μ, ν) with a change in the amplitude of the spectrum.

4. *Shift theorem*

$$\mathscr{F}\left[f\left(x-a, y-b\right)\right] = \exp\left[-\mathrm{i}2\pi\left(\mu a + \nu b\right)\right] F\left(\mu, \nu\right) . \tag{A.19}$$

Translation of a function in space coordinates introduces a linear phase shift in its spectrum.

5. *Parseval's theorem*

$$\iint_{-\infty}^{\infty} \left|f\left(x, y\right)\right|^2 \mathrm{d}x\mathrm{d}y = \iint_{-\infty}^{\infty} \left|F\left(\mu, \nu\right)\right|^2 \mathrm{d}\mu\mathrm{d}\nu . \tag{A.20}$$

This theorem presents the conservation of energy, that is, the total energy of a function in the space domain equals that of its spectrum in the frequency domain.

6. *Convolution theorem* The following mathematical operation of two functions $f(x, y)$ and $h(x, y)$

$$g\left(x, y\right) = \iint_{-\infty}^{\infty} f\left(\xi, \eta\right) h\left(x - \xi, y - \eta\right) \mathrm{d}\xi\mathrm{d}\eta \tag{A.21}$$

is generally called a two-dimensional convolution integral [168]. For this integral, we have the property expressed by

$$\mathscr{F}\left[\iint_{-\infty}^{\infty} f\left(\xi, \eta\right) h\left(x - \xi, y - \eta\right) \mathrm{d}\xi\mathrm{d}\eta\right]$$
$$= \mathscr{F}\left[\iint_{-\infty}^{\infty} h\left(\xi, \eta\right) f\left(x - \xi, y - \eta\right) \mathrm{d}\xi\mathrm{d}\eta\right]$$
$$= F\left(\mu, \nu\right) H\left(\mu, \nu\right) . \tag{A.22}$$

The convolution of two functions in the space domain corresponds to the simple multiplication of their spectra in the frequency domain. Similarly,

$$\mathscr{F}\left[f\left(x, y\right) h\left(x, y\right)\right] = \iint_{-\infty}^{\infty} F\left(\alpha, \beta\right) H\left(\mu - \alpha, \nu - \beta\right) \mathrm{d}\alpha\mathrm{d}\beta . \tag{A.23}$$

7. *Autocorrelation theorem* The cross-correlation of two functions $f(x, y)$ and $h(x, y)$ is defined by [168]

$$R_{\mathrm{c}}\left(x, y\right) = \iint_{-\infty}^{\infty} f\left(\xi, \eta\right) h^*\left(\xi - x, \eta - y\right) \mathrm{d}\xi\mathrm{d}\eta$$
$$= \iint_{-\infty}^{\infty} f\left(\xi + x, \eta + y\right) h^*\left(\xi, \eta\right) \mathrm{d}\xi\mathrm{d}\eta . \tag{A.24}$$

If $h(x, y) = f(x, y)$, the above equation is written as

$$R_{\mathrm{a}}\left(x, y\right) = \iint_{-\infty}^{\infty} f\left(\xi, \eta\right) f^*\left(\xi - x, \eta - y\right) \mathrm{d}\xi\mathrm{d}\eta$$
$$= \iint_{-\infty}^{\infty} f\left(\xi + x, \eta + y\right) f^*\left(\xi, \eta\right) \mathrm{d}\xi\mathrm{d}\eta . \tag{A.25}$$

This integral is called a two-dimensional autocorrelation, and the relation is as follows:

$$\mathscr{F}\left[\iint_{-\infty}^{\infty} f\left(\xi,\eta\right) f^{*}\left(\xi-x,\eta-y\right) \mathrm{d}\xi\mathrm{d}\eta\right] = \left|F\left(\mu,\nu\right)\right|^{2} . \tag{A.26}$$

The Fourier transform of an autocorrelation function results in a power spectrum, and this theorem is generally known as the Wiener–Khintchine theorem [168]. Similarly,

$$\mathscr{F}\left[\left|f\left(x,y\right)\right|^{2}\right] = \iint_{-\infty}^{\infty} F\left(\alpha,\beta\right) F^{*}\left(\alpha-\mu,\beta-\nu\right) \mathrm{d}\alpha\mathrm{d}\beta . \tag{A.27}$$

A.5 Examples of Fourier Transform Pairs

1. *Rectangle function and sinc function* For the rectangle function, rect(x), and the sinc function, sinc(x), defined by

$$\mathrm{rect}\left(x\right) = \begin{cases} 1, & -\dfrac{1}{2} \le x \le \dfrac{1}{2}, \\ 0, & \text{otherwise,} \end{cases} \tag{A.28}$$

and

$$\mathrm{sinc}\left(x\right) = \frac{\sin \pi x}{\pi x} , \tag{A.29}$$

we obtain the transform pairs

$$f\left(x,y\right) = \mathrm{rect}\left(x\right)\mathrm{rect}\left(y\right) , \tag{A.30}$$
$$\mathscr{F}\left[f\left(x,y\right)\right] = \mathrm{sinc}\left(\mu\right)\mathrm{sinc}\left(\nu\right) , \tag{A.31}$$

and

$$h\left(x,y\right) = \mathrm{sinc}\left(x\right)\mathrm{sinc}\left(y\right) , \tag{A.32}$$
$$\mathscr{F}\left[h\left(x,y\right)\right] = \mathrm{rect}\left(\mu\right)\mathrm{rect}\left(\nu\right) . \tag{A.33}$$

2. *Dirac delta function*

$$\delta\left(x,y\right) = \lim_{k\to\infty} k^{2} \exp\left[-k^{2}\pi\left(x^{2}+y^{2}\right)\right] , \tag{A.34}$$
$$\mathscr{F}\left[\delta\left(x,y\right)\right] = 1 , \tag{A.35}$$

and

$$f\left(x,y\right) = 1 , \tag{A.36}$$
$$\mathscr{F}\left[f\left(x,y\right)\right] = \delta\left(\mu,\nu\right) . \tag{A.37}$$

3. *Comb function* A comb function, $\mathrm{comb}(x)$, is defined by

$$\mathrm{comb}\,(x) = \sum_{n=-\infty}^{\infty} \delta\,(x-n)\;, \tag{A.38}$$

and for this function,

$$f\,(x,y) = \mathrm{comb}\,(x)\,\mathrm{comb}\,(y)\;, \tag{A.39}$$
$$\mathscr{F}\,[f\,(x,y)] = \mathrm{comb}\,(\mu)\,\mathrm{comb}\,(\nu)\;. \tag{A.40}$$

4. *Gauss function*

$$f\,(x,y) = \exp\left[-\pi\,\left(x^2 + y^2\right)\right]\;, \tag{A.41}$$
$$\mathscr{F}\,\{f\,(x,y)\} = \exp\left[-\pi\,\left(\mu^2 + \nu^2\right)\right]\;. \tag{A.42}$$

5. *Circle function* A circle function, $\mathrm{circ}(r)$, is defined by

$$\mathrm{circ}\,(r) = \begin{cases} 1\,, & r = \sqrt{x^2 + y^2} \le 1, \\ 0\,, & \text{otherwise,} \end{cases} \tag{A.43}$$

and for this function,

$$f\,(x,y) = \mathrm{circ}\,(r)\;, \tag{A.44}$$
$$\mathscr{F}\,[f\,(x,y)] = \frac{J_1\left(2\pi\sqrt{\mu^2 + \nu^2}\right)}{\sqrt{\mu^2 + \nu^2}}\;,$$
$$= \frac{J_1\,(2\pi\varrho)}{\varrho}\;, \tag{A.45}$$

where J_1 is a Bessel function of the first kind, the first order, and

$$\varrho = \sqrt{\mu^2 + \nu^2}\;. \tag{A.46}$$

B

Power Spectral Density of the Signal

The autocorrelation function $R(\tau_x, \tau_y)$ of the output signal $g(x_r, y_r)$ given in (2.1) is written by using formula (A.25) as

$$R(\tau_x, \tau_y) = \mathbf{E}\left[g(x_r + \tau_x, y_r + \tau_y)\, g^*(x_r, y_r)\right]$$

$$= \iint_{-\infty}^{\infty} g(x_r + \tau_x, y_r + \tau_y)\, g^*(x_r, y_r)\, \mathrm{d}x_r\, \mathrm{d}y_r \ . \tag{B.1}$$

Then, the power spectral density function $G_p(\mu, \nu)$ of $g(x_r, y_r)$ is obtained as the Fourier transform of $R(\tau_x, \tau_y)$ by using the Wiener–Khintchine theorem of (A.26):

$$G_p(\mu, \nu) = |G(\mu, \nu)|^2$$

$$= \iint_{-\infty}^{\infty} R(\tau_x, \tau_y)\, \exp\left[-\mathrm{i}2\pi\left(\mu\tau_x + \nu\tau_y\right)\right]\, \mathrm{d}\tau_x\, \mathrm{d}\tau_y \ . \tag{B.2}$$

By using (A.14), the Fourier spectrum $G(\mu, \nu)$ of the signal $g(x_r, y_r)$ can be expressed as

$$G(\mu, \nu) = \iint_{-\infty}^{\infty} g(x_r, y_r)\, \exp\left[-\mathrm{i}2\pi\left(\mu\tau_x + \nu\tau_y\right)\right]\, \mathrm{d}\tau_x\, \mathrm{d}\tau_y \ , \tag{B.3}$$

where the function $g(x_r, y_r)$ is given by the convolution integral of $f(x, y)$ and $h(x, y)$, as described in (2.1). By using the convolution theorem of (A.22), the Fourier spectrum $G(\mu, \nu)$ is written as

$$G(\mu, \nu) = F(\mu, \nu)\, H(\mu, \nu) \ , \tag{B.4}$$

where $F(\mu, \nu)$ and $H(\mu, \nu)$ are the Fourier spectra of $f(x, y)$ and $h(x, y)$, respectively. Then, the power spectrum $G_p(\mu, \nu)$ is expressed by

$$G_p(\mu, \nu) = |G(\mu, \nu)|^2$$

$$= |F(\mu, \nu)|^2\, |H(\mu, \nu)|^2$$

$$= F_p(\mu, \nu)\, H_p(\mu, \nu) \ , \tag{B.5}$$

where $F_{\mathrm{p}}(\mu, \nu)$ and $H_{\mathrm{p}}(\mu, \nu)$ are the power spectra of $f(x, y)$ and $h(x, y)$ given by (2.6) and (2.5), respectively. If $f(x, y)$ is a random function, it cannot be treated as a deterministic variable, and the power spectrum $F_{\mathrm{p}}(\mu, \nu)$ cannot be written in (2.6). What makes sense in this case is a statistical property such as an autocorrelation. The autocorrelation function $R_{\mathrm{F}}(\tau_{\mathrm{x}}, \tau_{\mathrm{y}})$ of the random process $f(x, y)$ is defined by [54]

$$
\begin{aligned}
R_{\mathrm{F}}(\tau_{\mathrm{x}}, \tau_{\mathrm{y}}) &= \mathbf{E}\left[f\left(x_{\mathrm{r}} + \tau_{\mathrm{x}}, y_{\mathrm{r}} + \tau_{\mathrm{y}}\right) f^{*}\left(x_{\mathrm{r}}, y_{\mathrm{r}}\right)\right] \\
&= \lim_{T_{\mathrm{x}}, T_{\mathrm{y}} \to \infty} \frac{1}{4T_{\mathrm{x}} T_{\mathrm{y}}} \\
&\quad \cdot \int_{-T_{\mathrm{x}}}^{T_{\mathrm{x}}} \int_{-T_{\mathrm{y}}}^{T_{\mathrm{y}}} f\left(x_{\mathrm{r}} + \tau_{\mathrm{x}}, y_{\mathrm{r}} + \tau_{\mathrm{y}}\right) f^{*}\left(x_{\mathrm{r}}, y_{\mathrm{r}}\right) \, \mathrm{d}x_{\mathrm{r}} \, \mathrm{d}y_{\mathrm{r}} \, . \quad \text{(B.6)}
\end{aligned}
$$

Then, the power spectrum $F_{\mathrm{p}}(\mu, \nu)$ is written, by using the Wiener–Khintchine theorem, as

$$
\begin{aligned}
F_{\mathrm{p}}(\mu, \nu) &= \iint_{-\infty}^{\infty} R_{\mathrm{F}}(\tau_{\mathrm{x}}, \tau_{\mathrm{y}}) \exp\left[-\mathrm{i}2\pi\left(\mu\tau_{\mathrm{x}} + \nu\tau_{\mathrm{y}}\right)\right] \mathrm{d}\tau_{\mathrm{x}} \, \mathrm{d}\tau_{\mathrm{y}} \\
&= \lim_{T_{\mathrm{x}}, T_{\mathrm{y}} \to \infty} \frac{1}{4T_{\mathrm{x}} T_{\mathrm{y}}} \\
&\quad \cdot \left| \int_{-T_{\mathrm{x}}}^{T_{\mathrm{x}}} \int_{-T_{\mathrm{y}}}^{T_{\mathrm{y}}} f(x, y) \exp\left[-\mathrm{i}2\pi\left(\mu x + \nu y\right)\right] \mathrm{d}x \, \mathrm{d}y \right|^{2} \, . \quad \text{(B.7)}
\end{aligned}
$$

C

Derivation of (2.12)

For the spatial filter shown in Fig. 2.3, the Fourier integral is carried out over the ranges of 0–X and 0–Y, instead of $-X/2$ to $+X/2$ and $-Y/2$ to $+Y/2$ for mathematical simplicity. This change introduces phase shifts but no influence on the result of the power spectrum $|H(\mu,\nu)|^2$. The Fourier spectrum $H(\mu,\nu)$ is calculated as

$$
\begin{aligned}
H(\mu,\nu) &= \iint_{-\infty}^{\infty} h(x,y) \exp\left[-i2\pi(\mu x + \nu y)\right] \, dx \, dy \\
&= \int_0^Y \exp(-i2\pi\nu y) \, dy \cdot \int_0^X h(x) \exp(-i2\pi\mu x) \, dx \\
&= \frac{1 - \exp(-i2\pi\nu Y)}{i2\pi\nu} H(\mu) \;,
\end{aligned}
\tag{C.1}
$$

where

$$
\begin{aligned}
H(\mu) &= \int_0^X h(x) \exp(-i2\pi\mu x) \, dx \\
&= \int_0^p h(x) \exp(-i2\pi\mu x) \, dx + \int_p^{2p} h(x) \exp(-i2\pi\mu x) \, dx \\
&\quad + \cdots + \int_{(n-1)p}^{np} h(x) \exp(-i2\pi\mu x) \, dx \;,
\end{aligned}
\tag{C.2}
$$

with the use of (2.11). By setting

$$
H_{s0}(\mu) = \int_0^p h(x) \exp(-i2\pi\mu x) \, dx \;,
\tag{C.3}
$$

(C.2) becomes

$$
\begin{aligned}
H(\mu) &= H_{s0}(\mu) + \left[\exp(-i2\pi\mu p)\right] H_{s0}(\mu) \\
&\quad + \left[\exp(-i2\pi\mu p)\right]^2 H_{s0}(\mu) + \cdots + \left[\exp(-i2\pi\mu p)\right]^{n-1} H_{s0}(\mu) \\
&= \frac{1 - \exp(-i2\pi\mu np)}{1 - \exp(-i2\pi\mu p)} H_{s0}(\mu) \;,
\end{aligned}
\tag{C.4}
$$

following the nature of a geometric series. Substitution of (C.4) in (C.1) with the use of

$$\sin \theta = \frac{\exp\left(i\theta\right) - \exp\left(-i\theta\right)}{i2} \tag{C.5}$$

yields

$$
\begin{aligned}
H\left(\mu, \nu\right) &= \frac{1 - \exp\left(-i2\pi\nu Y\right)}{i2\pi\nu} \cdot \frac{1 - \exp\left(-i2\pi\mu np\right)}{1 - \exp\left(-i2\pi\mu p\right)} H_{s0}\left(\mu\right) \\
&= \frac{\sin \pi\nu Y}{\pi\nu} \cdot \left[\exp\left(-i\pi\nu Y\right)\right] \cdot \frac{\sin \pi\mu np}{\sin \pi\mu p} \\
&\quad \cdot \left\{\exp\left[-i\pi\mu\left(n-1\right)p\right]\right\} H_{s0}\left(\mu\right) .
\end{aligned} \tag{C.6}
$$

Then, the power spectrum is expressed by

$$
\begin{aligned}
\left|H\left(\mu, \nu\right)\right|^2 &= H\left(\mu, \nu\right) H^*\left(\mu, \nu\right) \\
&= \left(\frac{\sin \pi\nu Y}{\pi\nu}\right)^2 \cdot \left(\frac{\sin \pi\mu np}{\pi\mu np}\right)^2 \cdot \frac{n^2}{\left(\frac{\sin \pi\mu p}{\pi\mu p}\right)^2} \cdot p^2 \left|\frac{1}{p} H_{s0}\left(\mu\right)\right|^2 \\
&= X^2 Y^2 \left(\frac{\sin \pi\nu Y}{\pi\nu Y}\right)^2 \cdot \left(\frac{\sin \pi\mu X}{\pi\mu X}\right)^2 \\
&\quad \cdot \left(\frac{\pi\mu p}{\sin \pi\mu p}\right)^2 \cdot \left|H_s\left(\mu\right)\right|^2 ,
\end{aligned} \tag{C.7}
$$

where $np = X$ was used and

$$H_s\left(\mu\right) = \frac{1}{p} \int_0^p h\left(x\right) \exp\left(-i2\pi\mu x\right) \, dx . \tag{C.8}$$

D

Derivation of (2.20) and (2.21)

For the sinusoidal transmittance $h_1(x)$, the integral is performed as follows:

$$
\begin{aligned}
H_{\mathrm{s1}}(\mu) &= \frac{1}{p}\int_0^p h_1(x)\exp(-\mathrm{i}2\pi\mu x)\,\mathrm{d}x \\
&= \frac{1}{2p}\int_0^p\left(1+\cos\frac{2\pi}{p}x\right)\exp(-\mathrm{i}2\pi\mu x)\,\mathrm{d}x \\
&= \frac{1}{2p}\int_0^p\exp(-\mathrm{i}2\pi\mu x)\,\mathrm{d}x + \frac{1}{4p}\int_0^p\exp\left[-\mathrm{i}2\pi\left(\mu-\frac{1}{p}\right)x\right]\,\mathrm{d}x \\
&\quad + \frac{1}{4p}\int_0^p\exp\left[-\mathrm{i}2\pi\left(\mu+\frac{1}{p}\right)x\right]\,\mathrm{d}x \\
&= \frac{1}{2p}\cdot\frac{\exp(-\mathrm{i}2\pi\mu p)-1}{-\mathrm{i}2\pi\mu} + \frac{1}{2p}\frac{\exp(-\mathrm{i}2\pi\mu p)\exp(\mathrm{i}2\pi)-1}{-\mathrm{i}4\pi\left(\mu-\dfrac{1}{p}\right)} \\
&\quad + \frac{1}{2p}\frac{\exp(-\mathrm{i}2\pi\mu p)\exp(-\mathrm{i}2\pi)-1}{-\mathrm{i}4\pi\left(\mu+\dfrac{1}{p}\right)} \\
&= \frac{\exp(-\mathrm{i}2\pi\mu p)-1}{2p}\left[\frac{1}{-\mathrm{i}2\pi\mu}+\frac{-\mathrm{i}2\pi\mu}{-(2\pi\mu)^2+\left(\dfrac{2\pi}{p}\right)^2}\right] \\
&= \frac{-\mathrm{i}\exp(-\mathrm{i}\pi\mu p)}{p}\cdot\sin(\pi\mu p)\cdot\frac{1}{-\mathrm{i}2\pi\mu}\cdot\frac{1}{1-\mu^2 p^2}\,,
\end{aligned}
\tag{D.1}
$$

where the relation

$$
\cos\theta = \frac{\exp(\mathrm{i}\theta)+\exp(-\mathrm{i}\theta)}{2}
\tag{D.2}
$$

and (C.5) are used. Note also that $|\exp(\mathrm{i}2\pi)| = |\exp(-\mathrm{i}2\pi)| = 1$. Then, the squared absolute of $H_{\mathrm{s1}}(\mu)$ is obtained as

$$|H_{s1}(\mu)|^2 = \left(\frac{\sin\pi\mu p}{\pi\mu p}\right)^2 \left[\frac{1}{2(1-\mu^2 p^2)}\right]^2 .$$

(D.3)

For the rectangular transmittance $h_2(x)$, the integral is simply performed over the range $-p/2$ to $+p/2$.

$$
\begin{aligned}
|H_{s2}(\mu)|^2 &= \left|\frac{1}{p}\int_{-\frac{p}{2}}^{\frac{p}{2}} h_2(x)\exp\left(-\mathrm{i}2\pi\mu x\right)\,\mathrm{d}x\right|^2 \\
&= \left|\frac{1}{p}\int_{-\frac{w}{2}}^{\frac{w}{2}} \exp\left(-\mathrm{i}2\pi\mu x\right)\,\mathrm{d}x\right|^2 \\
&= \left|\frac{1}{\pi\mu p}\cdot\frac{\exp\left(\mathrm{i}\pi\mu w\right)-\exp\left(-\mathrm{i}\pi\mu w\right)}{\mathrm{i}2}\right|^2 \\
&= \left(\frac{\sin\pi\mu w}{\pi\mu p}\right)^2 .
\end{aligned}
$$

(D.4)

D

Derivation of (2.20) and (2.21)

For the sinusoidal transmittance $h_1(x)$, the integral is performed as follows:

$$
\begin{aligned}
H_{s1}(\mu) &= \frac{1}{p}\int_0^p h_1(x)\exp\left(-i2\pi\mu x\right)\,\mathrm{d}x \\[2mm]
&= \frac{1}{2p}\int_0^p \left(1+\cos\frac{2\pi}{p}x\right)\exp\left(-i2\pi\mu x\right)\,\mathrm{d}x \\[2mm]
&= \frac{1}{2p}\int_0^p \exp\left(-i2\pi\mu x\right)\,\mathrm{d}x + \frac{1}{4p}\int_0^p \exp\left[-i2\pi\left(\mu-\frac{1}{p}\right)x\right]\,\mathrm{d}x \\[2mm]
&\quad + \frac{1}{4p}\int_0^p \exp\left[-i2\pi\left(\mu+\frac{1}{p}\right)x\right]\,\mathrm{d}x \\[2mm]
&= \frac{1}{2p}\cdot\frac{\exp\left(-i2\pi\mu p\right)-1}{-i2\pi\mu} + \frac{1}{2p}\frac{\exp\left(-i2\pi\mu p\right)\exp\left(i2\pi\right)-1}{-i4\pi\left(\mu-\dfrac{1}{p}\right)} \\[2mm]
&\quad + \frac{1}{2p}\frac{\exp\left(-i2\pi\mu p\right)\exp\left(-i2\pi\right)-1}{-i4\pi\left(\mu+\dfrac{1}{p}\right)} \\[2mm]
&= \frac{\exp\left(-i2\pi\mu p\right)-1}{2p}\left[\frac{1}{-i2\pi\mu} + \frac{-i2\pi\mu}{-(2\pi\mu)^2+\left(\dfrac{2\pi}{p}\right)^2}\right] \\[2mm]
&= \frac{-i\exp\left(-i\pi\mu p\right)}{p}\cdot\sin\left(\pi\mu p\right)\cdot\frac{1}{-i2\pi\mu}\cdot\frac{1}{1-\mu^2 p^2}\,, \tag{D.1}
\end{aligned}
$$

where the relation

$$
\cos\theta = \frac{\exp\left(i\theta\right)+\exp\left(-i\theta\right)}{2} \tag{D.2}
$$

and (C.5) are used. Note also that $|\exp(i2\pi)| = |\exp(-i2\pi)| = 1$. Then, the squared absolute of $H_{s1}(\mu)$ is obtained as

$$|H_{s1}(\mu)|^2 = \left(\frac{\sin\pi\mu p}{\pi\mu p}\right)^2 \left[\frac{1}{2(1-\mu^2 p^2)}\right]^2 . \tag{D.3}$$

For the rectangular transmittance $h_2(x)$, the integral is simply performed over the range $-p/2$ to $+p/2$.

$$\begin{aligned}
|H_{s2}(\mu)|^2 &= \left|\frac{1}{p}\int_{-\frac{p}{2}}^{\frac{p}{2}} h_2(x)\exp\left(-\mathrm{i}2\pi\mu x\right)\,\mathrm{d}x\right|^2 \\[2mm]
&= \left|\frac{1}{p}\int_{-\frac{w}{2}}^{\frac{w}{2}} \exp\left(-\mathrm{i}2\pi\mu x\right)\,\mathrm{d}x\right|^2 \\[2mm]
&= \left|\frac{1}{\pi\mu p}\cdot\frac{\exp\left(\mathrm{i}\pi\mu w\right)-\exp\left(-\mathrm{i}\pi\mu w\right)}{\mathrm{i}2}\right|^2 \\[2mm]
&= \left(\frac{\sin\pi\mu w}{\pi\mu p}\right)^2 .
\end{aligned} \tag{D.4}$$

E

Power Spectra for Spatial Filters in Sect. 2.3

E.1 Derivation of (2.24)

$$H\left(\mu,\nu\right) = \int_{-\frac{X}{2}}^{\frac{X}{2}} \int_{-\frac{Y}{2}}^{\frac{Y}{2}} \frac{1}{2}\left(1+\cos\frac{2\pi}{p}x\right)\exp\left[-\mathrm{i}2\pi\left(\mu x+\nu y\right)\right]\,\mathrm{d}x\,\mathrm{d}y$$

$$= \frac{1}{2}I_{\mathrm{y}}\left(I_{\mathrm{x1}}+I_{\mathrm{x2}}\right),\tag{E.1}$$

where

$$I_{\mathrm{y}} = \int_{-\frac{Y}{2}}^{\frac{Y}{2}} \exp\left(-\mathrm{i}2\pi\nu y\right)\,\mathrm{d}y = Y\cdot\frac{\sin\pi\nu Y}{\pi\nu Y},\tag{E.2}$$

and

$$I_{\mathrm{x1}} = \int_{-\frac{X}{2}}^{\frac{X}{2}} \exp\left(-\mathrm{i}2\pi\mu x\right)\,\mathrm{d}x = X\cdot\frac{\sin\pi\mu X}{\pi\mu X},\tag{E.3}$$

$$I_{\mathrm{x2}} = \int_{-\frac{X}{2}}^{\frac{X}{2}} \cos\frac{2\pi}{p}x\cdot\exp\left(-\mathrm{i}2\pi\mu x\right)\,\mathrm{d}x$$

$$= \frac{1}{2}\int_{-\frac{X}{2}}^{\frac{X}{2}}\left\{\exp\left[-\mathrm{i}2\pi\left(\mu-\frac{1}{p}\right)x\right] - \exp\left[-\mathrm{i}2\pi\left(\mu+\frac{1}{p}\right)x\right]\right\}\,\mathrm{d}x$$

$$= \frac{X}{2}\left[\frac{\sin\pi\left(\mu-\frac{1}{p}\right)X}{\pi\left(\mu-\frac{1}{p}\right)X} + \frac{\sin\pi\left(\mu+\frac{1}{p}\right)X}{\pi\left(\mu+\frac{1}{p}\right)X}\right].\tag{E.4}$$

Substitution of (E.2)–(E.4) in (E.1) and taking the squared absolute yield

$$\left|H\left(\mu,\nu\right)\right|^2 = \frac{X^2Y^2}{4}\left|H_{\mathrm{Y}}\left(\nu\right)\right|^2\left\{\left[H_{\mathrm{X}}\left(\mu\right)+\frac{1}{2}\left[H_{\mathrm{X}}^{-}\left(\mu\right)+H_{\mathrm{X}}^{+}\left(\mu\right)\right]\right\}^2,\tag{E.5}$$

where

$$H_Y(\nu) = \frac{\sin \pi \nu Y}{\pi \nu Y} \, ,$$

$$H_X(\mu) = \frac{\sin \pi \mu X}{\pi \mu X} \, ,$$

$$H_X^-(\mu) = \frac{\sin \pi \left(\mu - \dfrac{1}{p} \right) X}{\pi \left(\mu - \dfrac{1}{p} \right) X} \, ,$$

$$H_X^+(\mu) = \frac{\sin \pi \left(\mu + \dfrac{1}{p} \right) X}{\pi \left(\mu + \dfrac{1}{p} \right) X} \, .$$

E.2 Derivation of (2.30)

$$H(\mu, \nu) = \iint_{x^2 + y^2 \leq a^2} \frac{1}{2} \left(1 + \cos \frac{2\pi}{p} x \right)$$
$$\cdot \exp\left[-i2\pi (\mu x + \nu y) \right] \, dx \, dy \, . \tag{E.6}$$

By setting $x = r \cos \phi$ and $y = r \sin \phi$ in polar coordinates (r, ϕ), (E.6) is written as

$$H(\mu, \nu) = \frac{1}{2} \int_0^a \int_0^{2\pi} \exp\left[-i2\pi r (\mu \cos \phi + \nu \sin \phi) \right] r \, dr \, d\phi$$
$$+ \frac{1}{2} \int_0^a \int_0^{2\pi} \cos \left(\frac{2\pi}{p} r \cos \phi \right)$$
$$\cdot \exp\left[-i2\pi r (\mu \cos \phi + \nu \sin \phi) \right] r \, dr \, d\phi$$
$$= I_1 + I_2 \, . \tag{E.7}$$

Setting $\psi = \tan^{-1}(\mu / \nu)$ gives the transformation,

$$\mu \cos \phi + \nu \sin \phi = \sqrt{\mu^2 + \nu^2} \sin (\phi + \psi) = \varrho \sin \phi' \, , \tag{E.8}$$

where $\varrho = \sqrt{\mu^2 + \nu^2}$ and $\phi' = \phi + \psi$. By using the two Bessel-function identities,

$$J_0(u) = \frac{1}{2\pi} \int_0^{2\pi} \exp\left(-iu \sin \theta \right) d\theta \, , \tag{E.9}$$

$$\varrho J_1(\varrho) = \int_0^{\varrho} u J_0(u) \, du \, , \tag{E.10}$$

where $J_0(u)$ and $J_1(\varrho)$ are Bessel functions of the first kind, zero and first orders, respectively, the following derivations are obtained:

$$I_1 = \frac{1}{2} \int_0^a \int_\psi^{2\pi+\psi} \exp\left[-\mathrm{i}2\pi r\varrho \sin\phi'\right] r \, \mathrm{d}r \, \mathrm{d}\phi'$$

$$= \pi \int_0^a J_0\left(2\pi r\varrho\right) r \, \mathrm{d}r$$

$$= \pi a^2 \frac{J_1\left(2\pi a\varrho\right)}{2\pi a\varrho} \,. \tag{E.11}$$

$$I_2 = \frac{1}{4} \int_0^a \int_0^{2\pi} \left[\exp\left(\mathrm{i}\frac{2\pi}{p} r \cos\phi\right) + \exp\left(-\mathrm{i}\frac{2\pi}{p} r \cos\phi\right)\right]$$
$$\cdot \exp\left[-\mathrm{i}2\pi r\varrho \sin(\phi+\psi)\right] r \, \mathrm{d}r \, \mathrm{d}\phi$$

$$= \frac{1}{4} \int_0^a \int_0^{2\pi} \exp\left\{-\mathrm{i}2\pi r\left[\varrho \sin(\phi+\psi) - \frac{1}{p}\cos\phi\right]\right\} r \, \mathrm{d}r \, \mathrm{d}\phi$$

$$+ \frac{1}{4} \int_0^a \int_0^{2\pi} \exp\left\{-\mathrm{i}2\pi r\left[\varrho \sin(\phi+\psi) + \frac{1}{p}\cos\phi\right]\right\} r \, \mathrm{d}r \, \mathrm{d}\phi$$

$$= I_{21} + I_{22} \,. \tag{E.12}$$

$$I_{21} = \frac{1}{4} \int_0^a \int_0^{2\pi} \exp\left[-\mathrm{i}2\pi rC_1 \sin(\phi+\Phi_1)\right] r \, \mathrm{d}r \, \mathrm{d}\phi \,, \tag{E.13}$$

where

$$C_1 = \sqrt{(\varrho\cos\psi)^2 + \left(\varrho\sin\psi - \frac{1}{p}\right)^2} = \sqrt{\left(\mu - \frac{1}{p}\right)^2 + \nu^2} \tag{E.14}$$

with the use of the relations $\mu = \varrho\sin\psi$ and $\nu = \varrho\cos\psi$, and

$$\Phi_1 = \tan^{-1}\left\{\frac{\varrho\sin\psi - (1/p)}{\varrho\cos\psi}\right\} \,. \tag{E.15}$$

In the same way as I_{21},

$$I_{22} = \frac{1}{4} \int_0^a \int_0^{2\pi} \exp\left[-\mathrm{i}2\pi rC_2 \sin(\phi+\Phi_2)\right] r \, \mathrm{d}r \, \mathrm{d}\phi \,, \tag{E.16}$$

$$C_2 = \sqrt{\left(\mu + \frac{1}{p}\right)^2 + \nu^2} \,, \tag{E.17}$$

$$\Phi_2 = \tan^{-1}\left[\frac{\varrho\sin\psi + (1/p)}{\varrho\cos\psi}\right] \,. \tag{E.18}$$

Then,

$$I_2 = \frac{1}{4} \int_0^a \int_{\Phi_1}^{2\pi+\Phi_1} \exp\left(-\mathrm{i}2\pi rC_1 \sin\alpha_1\right) r \, \mathrm{d}r \, \mathrm{d}\alpha_1$$

$$+ \frac{1}{4} \int_0^a \int_{\Phi_2}^{2\pi + \Phi_2} \exp\left(-\mathrm{i}2\pi r C_2 \sin \alpha_2\right) r \, \mathrm{d}r \, \mathrm{d}\alpha_2$$

$$= \frac{\pi}{2} \int_0^a J_0\left(2\pi r C_1\right) r \, \mathrm{d}r + \frac{\pi}{2} \int_0^a J_0\left(2\pi r C_2\right) r \, \mathrm{d}r$$

$$= \frac{\pi}{2} \cdot \frac{a}{2\pi C_1} J_1\left(2\pi a C_1\right) + \frac{\pi}{2} \cdot \frac{a}{2\pi C_2} J_1\left(2\pi a C_2\right) , \qquad (\text{E.19})$$

where $\alpha_1 = \phi + \Phi_1$ and $\alpha_2 = \phi + \Phi_2$. Finally, the following equation is obtained;

$$|H\left(\mu, \nu\right)|^2 = |I_1 + I_2|^2$$

$$= \pi^2 a^4 \left\{ H_J\left(\mu, \nu\right) + \frac{1}{2}\left[H_J^-\left(\mu, \nu\right) + H_J^+\left(\mu, \nu\right)\right] \right\}^2 , \qquad (\text{E.20})$$

where

$$H_J\left(\mu, \nu\right) = \frac{J_1\left(2\pi a \sqrt{\mu^2 + \nu^2}\right)}{2\pi a \sqrt{\mu^2 + \nu^2}} ,$$

$$H_J^-\left(\mu, \nu\right) = \frac{J_1\left[2\pi a \sqrt{\left(\mu - \frac{1}{p}\right)^2 + \nu^2}\right]}{2\pi a \sqrt{\left(\mu - \frac{1}{p}\right)^2 + \nu^2}} ,$$

$$H_J^+\left(\mu, \nu\right) = \frac{J_1\left[2\pi a \sqrt{\left(\mu + \frac{1}{p}\right)^2 + \nu^2}\right]}{2\pi a \sqrt{\left(\mu + \frac{1}{p}\right)^2 + \nu^2}} ,$$

E.3 Derivation of (2.34)

$$H\left(\mu, \nu\right) = \iint_{-\infty}^{\infty} \frac{1}{2} \exp\left(-\frac{x^2 + y^2}{2\sigma^2}\right) \cdot \left(1 + \cos \frac{2\pi}{p} x\right)$$

$$\cdot \exp\left[-\mathrm{i}2\pi\left(\mu x + \nu y\right)\right] \, \mathrm{d}x \, \mathrm{d}y$$

$$= \frac{1}{2} I_y\left(I_{x1} + I_{x2}\right) , \qquad (\text{E.21})$$

where

$$I_y = \int_{-\infty}^{\infty} \exp\left(-\frac{y^2}{2\sigma^2}\right) \cdot \exp\left(-\mathrm{i}2\pi \nu y\right) \, \mathrm{d}y ,$$

$$I_{\mathrm{x1}} = \int_{-\infty}^{\infty} \exp\left(-\frac{x^2}{2\sigma^2}\right) \cdot \exp\left(-\mathrm{i}2\pi\mu x\right)\,\mathrm{d}x \;,$$

$$I_{\mathrm{x2}} = \frac{1}{2} \int_{-\infty}^{\infty} \left[\exp\left(\mathrm{i}\frac{2\pi}{p}x\right) + \exp\left(-\mathrm{i}\frac{2\pi}{p}x\right)\right]$$

$$\cdot \exp\left(-\frac{x^2}{2\sigma^2}\right) \exp\left(-\mathrm{i}2\pi\mu x\right)\,\mathrm{d}x \;.$$

Note that the three integrals I_{y}, I_{x1}, and I_{x2} above are the Fourier transforms of Gaussian functions and, then,

$$I_{\mathrm{y}} \;=\; \sqrt{2\pi}\sigma \exp\left(-2\pi^2\sigma^2\nu^2\right) \;, \tag{E.22}$$

$$I_{\mathrm{x1}} \;=\; \sqrt{2\pi}\sigma \exp\left(-2\pi^2\sigma^2\mu^2\right) \;, \tag{E.23}$$

$$I_{\mathrm{x2}} = \frac{1}{2} \int_{-\infty}^{\infty} \exp\left(-\frac{x^2}{2\sigma^2}\right) \exp\left[-\mathrm{i}2\pi\left(\mu - \frac{1}{p}\right)x\right]\,\mathrm{d}x$$

$$+ \frac{1}{2} \int_{-\infty}^{\infty} \exp\left(-\frac{x^2}{2\sigma^2}\right) \exp\left[-\mathrm{i}2\pi\left(\mu + \frac{1}{p}\right)x\right]\,\mathrm{d}x$$

$$= \sqrt{\frac{\pi}{2}}\sigma \left\{ \exp\left[-2\pi^2\sigma^2\left(\mu - \frac{1}{p}\right)^2\right] + \exp\left[-2\pi^2\sigma^2\left(\mu + \frac{1}{p}\right)^2\right]\right\} \;. \tag{E.24}$$

Finally, the power spectrum is obtained as

$$|H\left(\mu,\nu\right)|^2 = \pi^2\sigma^4 \left\{ H_{\mathrm{G}}\left(\mu,\nu\right) + \frac{1}{2}\left[H_{\mathrm{G}}^{-}\left(\mu,\nu\right) + H_{\mathrm{G}}^{+}\left(\mu,\nu\right)\right]\right\}^2 \;, \tag{E.25}$$

where

$$H_{\mathrm{G}}\left(\mu,\nu\right) = \exp\left[-2\pi^2\sigma^2\left(\mu^2 + \nu^2\right)\right] \;,$$

$$H_{\mathrm{G}}^{-}\left(\mu,\nu\right) = \exp\left\{-2\pi^2\sigma^2\left[\left(\mu - \frac{1}{p}\right)^2 + \nu^2\right]\right\} \;,$$

$$H_{\mathrm{G}}^{+}\left(\mu,\nu\right) = \exp\left\{-2\pi^2\sigma^2\left[\left(\mu + \frac{1}{p}\right)^2 + \nu^2\right]\right\} \;.$$

F

Derivation of (2.45)

The rotation by an angle θ in the moving direction of the object in Fig. 2.16 can be treated equivalently by rotation of the spatial filter by the same angle for mathematical simplicity. In this case, the function $h(x, y)$ for the circular type with sinusoidal transmittance is written, by using (2.44), as

$$
h\left(x, y\right) = \begin{cases} \dfrac{1}{2}\left\{1 + \cos\left[\dfrac{2\pi}{p}\left(x\cos\theta + y\sin\theta\right)\right]\right\}, & x^2 + y^2 \le a^2, \\ 0, & \text{otherwise.} \end{cases} \tag{F.1}
$$

The power spectrum is, then, written in polar coordinates (r, ϕ) as

$$
\begin{aligned}
H\left(\mu, \nu\right) &= \frac{1}{2}\int_0^a\int_0^{2\pi}\left\{1 + \cos\left[\frac{2\pi r}{p}\left(\cos\phi\cos\theta + \sin\phi\sin\theta\right)\right]\right\} \\
&\quad \cdot \exp\left[-\mathrm{i}2\pi r\left(\mu\cos\phi + \nu\sin\phi\right)\right] r\,\mathrm{d}r\,\mathrm{d}\phi \\
&= I_1 + I_2^+ + I_2^- ,
\end{aligned} \tag{F.2}
$$

where

$$
I_1 = \frac{1}{2}\int_0^a\int_0^{2\pi}\exp\left[-\mathrm{i}2\pi r\left(\mu\cos\phi + \nu\sin\phi\right)\right] r\,\mathrm{d}r\,\mathrm{d}\phi ,
$$

$$
I_2^+ = \frac{1}{4}\int_0^a\int_0^{2\pi} E^+\exp\left[-\mathrm{i}2\pi r\sqrt{\mu^2 + \nu^2}\right]\sin\left(\phi + \psi\right) r\,\mathrm{d}r\,\mathrm{d}\phi ,
$$

$$
I_2^- = \frac{1}{4}\int_0^a\int_0^{2\pi} E^-\exp\left[-\mathrm{i}2\pi r\sqrt{\mu^2 + \nu^2}\right]\sin\left(\phi + \psi\right) r\,\mathrm{d}r\,\mathrm{d}\phi ,
$$

and

$$
E^+ = \exp\left[\mathrm{i}\frac{2\pi r}{p}\left(\cos\phi\cos\theta + \sin\phi\sin\theta\right)\right] ,
$$

$$
E^- = \exp\left[-\mathrm{i}\frac{2\pi r}{p}\left(\cos\phi\cos\theta + \sin\phi\sin\theta\right)\right] .
$$

In the above derivation, (E.8) has been used. The integral I_1 above is identical to I_1 in (E.7) and, thus,

$$I_1 = \pi a^2 \frac{J_1\left(2\pi a\sqrt{\mu^2 + \nu^2}\right)}{2\pi a\sqrt{\mu^2 + \nu^2}} \ . \tag{F.3}$$

On the other hand,

$$\begin{aligned}
I_2^+ &= \frac{1}{4}\int_0^a\int_0^{2\pi} \exp\left\{-\mathrm{i}2\pi r\left[\varrho\sin(\phi+\psi) - \frac{1}{p}\cos(\phi-\theta)\right]\right\} r\,\mathrm{d}r\,\mathrm{d}\phi \\
&= \frac{1}{4}\int_0^a\int_0^{2\pi} \exp\left[-\mathrm{i}2\pi r\sqrt{A^2 + B^2}\sin(\phi+\Phi)\right] r\,\mathrm{d}r\,\mathrm{d}\phi \\
&= \frac{\pi a^2}{4}\cdot\frac{2J_1\left(2\pi a\sqrt{A^2 + B^2}\right)}{2\pi a\sqrt{A^2 + B^2}} \ ,
\end{aligned} \tag{F.4}$$

and

$$\begin{aligned}
I_2^- &= \frac{1}{4}\int_0^a\int_0^{2\pi} \exp\left\{-\mathrm{i}2\pi r\left[\varrho\sin(\phi+\psi) + \frac{1}{p}\cos(\phi-\theta)\right]\right\} r\,\mathrm{d}r\,\mathrm{d}\phi \\
&= \frac{1}{4}\int_0^a\int_0^{2\pi} \exp\left[-\mathrm{i}2\pi r\sqrt{A'^2 + B'^2}\sin(\phi+\Phi')\right] r\,\mathrm{d}r\,\mathrm{d}\phi \\
&= \frac{\pi a^2}{4}\cdot\frac{2J_1\left(2\pi a\sqrt{A'^2 + B'^2}\right)}{2\pi a\sqrt{A'^2 + B'^2}} \ .
\end{aligned} \tag{F.5}$$

Thus,

$$I_2^+ + I_2^- = \frac{\pi a^2}{4}\left[\frac{2J_1\left(2\pi a\sqrt{A^2 + B^2}\right)}{2\pi a\sqrt{A^2 + B^2}} + \frac{2J_1\left(2\pi a\sqrt{A'^2 + B'^2}\right)}{2\pi a\sqrt{A'^2 + B'^2}}\right] \ , \tag{F.6}$$

where

$$A = \varrho\cos\psi - \frac{1}{p}\sin\theta \ , \tag{F.7}$$

$$A' = \varrho\cos\psi + \frac{1}{p}\sin\theta \ , \tag{F.8}$$

$$B = \varrho\sin\psi - \frac{1}{p}\cos\theta \ , \tag{F.9}$$

$$B' = \varrho\sin\psi + \frac{1}{p}\cos\theta \ , \tag{F.10}$$

$$\varrho = \sqrt{\mu^2 + \nu^2} \ ,$$

$$\Phi = \tan^{-1}\frac{B}{A} \ ,$$

$$\Phi' = \tan^{-1}\frac{B'}{A'} \ .$$

Use of (F.7)–(F.10) yields

$$A^2 + B^2 \;=\; \left(\mu - \frac{\cos\theta}{p}\right)^2 + \left(\nu - \frac{\sin\theta}{p}\right)^2 , \tag{F.11}$$

$$A'^2 + B'^2 \;=\; \left(\mu + \frac{\cos\theta}{p}\right)^2 + \left(\nu + \frac{\sin\theta}{p}\right)^2 . \tag{F.12}$$

Finally, the power spectrum is derived as

$$|H(\mu,\nu)|^2 = \pi^2 a^4 \left\{ H_{\mathrm{J}}(\mu,\nu) + \frac{1}{2}\left[H^{-}_{\mathrm{J}\theta}(\mu,\nu) + H^{+}_{\mathrm{J}\theta}(\mu,\nu)\right]\right\}^2 , \tag{F.13}$$

where

$$H_{\mathrm{J}}(\mu,\nu) \;=\; \frac{J_1\left(2\pi a\sqrt{\mu^2+\nu^2}\right)}{2\pi a\sqrt{\mu^2+\nu^2}} ,$$

$$H^{-}_{\mathrm{J}\theta}(\mu,\nu) = \frac{J_1\left[2\pi a\sqrt{\left(\mu - \dfrac{\cos\theta}{p}\right)^2 + \left(\nu - \dfrac{\sin\theta}{p}\right)^2}\right]}{2\pi a\sqrt{\left(\mu - \dfrac{\cos\theta}{p}\right)^2 + \left(\nu - \dfrac{\sin\theta}{p}\right)^2}} ,$$

$$H^{+}_{\mathrm{J}\theta}(\mu,\nu) = \frac{J_1\left[2\pi a\sqrt{\left(\mu + \dfrac{\cos\theta}{p}\right)^2 + \left(\nu + \dfrac{\sin\theta}{p}\right)^2}\right]}{2\pi a\sqrt{\left(\mu + \dfrac{\cos\theta}{p}\right)^2 + \left(\nu + \dfrac{\sin\theta}{p}\right)^2}} .$$

G

One-Dimensional Power Spectrum of the Signal

With reference to the derivation of (2.34) in Appendix E.3, the one-dimensional power spectrum $H_\mathrm{p}(\mu) = |H(\mu)|^2$ of the spatial filter is derived as

$$H_\mathrm{p}\left(\mu\right) = \frac{\pi\sigma^2}{2}\left\{H_\mathrm{G}\left(\mu\right) + \frac{1}{2}\left[H_\mathrm{G}^-\left(\mu\right) + H_\mathrm{G}^+\left(\mu\right)\right]\right\}^2 , \qquad (\mathrm{G.1})$$

where

$$H_\mathrm{G}\left(\mu\right) = \exp\left(-2\pi^2\sigma^2\mu^2\right) ,$$

$$H_\mathrm{G}^-\left(\mu\right) = \exp\left[-2\pi^2\sigma^2\left(\mu - \frac{1}{p}\right)^2\right] ,$$

$$H_\mathrm{G}^+\left(\mu\right) = \exp\left[-2\pi^2\sigma^2\left(\mu + \frac{1}{p}\right)^2\right] .$$

Then, the power spectrum $G_\mathrm{p}(\mu)$ is derived with (2.50) as

$$\begin{aligned}
G_\mathrm{p}(\mu) &= H_\mathrm{p}(\mu)F_\mathrm{p}(\mu) \\
&= \frac{\pi\sigma^2}{2}\left[G_1 + G_2 + G_3\right] , \qquad (\mathrm{G.2})
\end{aligned}$$

where

$$G_1 = \left[1 + \frac{1}{2}\exp\left(-\frac{4\pi^2\sigma^2}{p^2}\right)\right]\exp\left(-c^2\mu^2\right) ,$$

$$G_2 = E_2\left\{\exp\left[-c^2\left(\mu - \frac{d^2}{2p}\right)^2\right] + \exp\left[-c^2\left(\mu + \frac{d^2}{2p}\right)^2\right]\right\} ,$$

$$G_3 = E_3\left\{\exp\left[-c^2\left(\mu - \frac{d^2}{p}\right)^2\right] + \exp\left[-c^2\left(\mu + \frac{d^2}{p}\right)^2\right]\right\} ,$$

and

$$E_2 = \exp\left[-\frac{\pi^2\sigma^2}{p^2}\left(2 - d^2\right)\right] ,$$

$$E_3 = \frac{1}{4}\exp\left[-\frac{4\pi^2\sigma^2}{p^2}\left(1 - d^2\right)\right] ,$$

$$c^2 = 4\pi^2\left(b_{\mathrm{g}}^2 + \sigma^2\right) , \tag{G.3}$$

$$d^2 = \frac{\sigma^2}{b_{\mathrm{g}}^2 + \sigma^2} . \tag{G.4}$$

In (G.2), the following two terms become sufficiently small in comparison with unity for usual spatial filters:

$$\frac{1}{2}\exp\left(-\frac{4\pi^2\sigma^2}{p^2}\right)$$

and

$$\exp\left[-\frac{\pi^2\sigma^2}{p^2}\left(2 - d^2\right)\right] .$$

Then, $G_{\mathrm{p}}(\mu)$ is derived approximately as

$$G_{\mathrm{p}}(\mu) \simeq \frac{\pi\sigma^2}{2}\exp\left[-4\pi^2\left(b_{\mathrm{g}}^2 + \sigma^2\right)\mu^2\right]$$
$$+ \frac{\pi\sigma^2}{8}\exp\left[-\frac{4\pi^2\sigma^2}{p^2}\left(1 - \frac{\sigma^2}{b_{\mathrm{g}}^2 + \sigma^2}\right)\right]\left(G_{\mathrm{p}}^- + G_{\mathrm{p}}^+\right) , \tag{G.5}$$

where

$$G_{\mathrm{p}}^- = \exp\left[-4\pi^2\left(b_{\mathrm{g}}^2 + \sigma^2\right)\left(\mu - \frac{1}{p}\cdot\frac{\sigma^2}{b_{\mathrm{g}}^2 + \sigma^2}\right)^2\right] ,$$

$$G_{\mathrm{p}}^+ = \exp\left[-4\pi^2\left(b_{\mathrm{g}}^2 + \sigma^2\right)\left(\mu + \frac{1}{p}\cdot\frac{\sigma^2}{b_{\mathrm{g}}^2 + \sigma^2}\right)^2\right] .$$

The central frequency of the signal component is obtained from the above equation as

$$\mu = \pm\frac{1}{p}\cdot\frac{\sigma^2}{b_{\mathrm{g}}^2 + \sigma^2} = \pm\frac{1}{p}\cdot\frac{\dfrac{4\sigma^2}{p^2}}{\dfrac{4b_{\mathrm{g}}^2}{p^2} + \dfrac{4\sigma^2}{p^2}}$$

$$= \pm\frac{1}{p}\cdot\frac{n^2}{\left(\dfrac{2b_{\mathrm{g}}}{p}\right)^2 + n^2} . \tag{G.6}$$

Derivation of Output Signals for Visibility Analysis

For mathematical simplicity, the convolution integral of (2.1) is performed in a modified manner as

$$g\left(t\right) = \iint_{-\infty}^{\infty} f\left(x, y\right) h\left(x - v_{\mathrm{x}} t, y\right) \, \mathrm{d}x \, \mathrm{d}y \, . \tag{H.1}$$

This modification makes no change in the results in the expression of output signals due to the relative displacement between $f(x, y)$ and $h(x, y)$ in the convolution integral.

H.1 Derivation of (2.55)

Use of (2.53) and (2.54) in (H.1) and conversion of the result into polar coordinates (r, ϕ) yield

$$\begin{aligned}
g\left(t\right) &= \iint_{x^2 + y^2 \leq b^2} \frac{I_0}{2} \left\{ 1 + \cos\left[\frac{2\pi}{p}\left(x - v_{\mathrm{x}} t\right)\right] \right\} \mathrm{d}x \, \mathrm{d}y \\
&= \frac{I_0}{2} \int_0^b \int_0^{2\pi} r \, \mathrm{d}r \, \mathrm{d}\phi \\
&\quad + \frac{I_0}{2} \int_0^b \int_0^{2\pi} r \cos\left[\frac{2\pi}{p}\left(r \cos\phi - v_{\mathrm{x}} t\right)\right] \mathrm{d}r \, \mathrm{d}\phi \\
&= I_1 + I_2 \, ,
\end{aligned} \tag{H.2}$$

where

$$I_1 = \frac{I_0}{2} \int_0^b \int_0^{2\pi} r \, \mathrm{d}r \, \mathrm{d}\phi = \frac{I_0 b^2}{4} \int_0^{2\pi} \mathrm{d}\phi = \frac{I_0 \pi b^2}{2} \, , \tag{H.3}$$

and

$$\begin{aligned}
I_2 &= \frac{I_0}{2} \int_0^b \int_0^{2\pi} r \cos\left[\frac{2\pi}{p}\left(r \cos\phi - v_{\mathrm{x}} t\right)\right] r \, \mathrm{d}r \, \mathrm{d}\phi \\
&= I_{2\mathrm{c}} + I_{2\mathrm{s}} \, ,
\end{aligned} \tag{H.4}$$

where

$$I_{2c} = \frac{I_0}{2} \cos\left(\frac{2\pi v_x}{p} t\right) \cdot \int_0^b \int_0^{2\pi} r \cos\left(\frac{2\pi}{p} r \cos\phi\right) \, dr \, d\phi \,,$$

$$I_{2s} = \frac{I_0}{2} \sin\left(\frac{2\pi v_x}{p} t\right) \cdot \int_0^b \int_0^{2\pi} r \sin\left(\frac{2\pi}{p} r \cos\phi\right) \, dr \, d\phi \,.$$

With the use of the Bessel function's formulas

$$\int_0^{2\pi} \cos\left(t \cos\phi\right) \, d\phi = 2\pi J_0\left(t\right) \,, \tag{H.5}$$

$$\int_0^{2\pi} \sin\left(t \cos\phi\right) \, d\phi = 0 \,, \tag{H.6}$$

the integral I_2 is written as

$$I_2 = \frac{I_0}{2} \cos\left(\frac{2\pi v_x}{p} t\right) \cdot \int_0^b 2\pi r J_0\left(\frac{2\pi r}{p}\right) \, dr + 0$$

$$= \frac{I_0}{2} \cos\left(\frac{2\pi v_x}{p} t\right) \cdot 2\pi b^2 \frac{J_1\left(\frac{2\pi b}{p}\right)}{\frac{2\pi b}{p}} \,, \tag{H.7}$$

where the relation of (E.10) has been used. Then, the output signal is derived as

$$g\left(t\right) = \frac{I_0 \pi b^2}{2} \left[1 + \frac{2 J_1\left(\frac{2\pi b}{p}\right)}{\frac{2\pi b}{p}} \cos\left(\frac{2\pi v_x}{p} t\right)\right] \,. \tag{H.8}$$

For a rectangular-transmittance filter, the same procedure as above can be applied to each of the fundamental and higher harmonic cosine functions of $h(x,y)$;

$$h\left(x,y\right) = \frac{1}{2} + \sum_{m=1}^{\infty} (-1)^{m-1} \frac{2}{(2m-1)\pi} \cos\left[\frac{2\left(2m-1\right)\pi}{p} x\right] \,. \tag{H.9}$$

Then, (2.57) is simply derived.

H.2 Derivation of (2.59)

Using the expression for $f(x,y)$ as

$$f\left(x,y\right) = I_0 \exp\left(-\frac{x^2 + y^2}{2 b_g^2}\right) \,, \tag{H.10}$$

$$g\left(t\right) = \frac{I_0}{2} \iint_{-\infty}^{\infty} \left\{1 + \cos\left[\frac{2\pi}{p}\left(x - v_{\mathrm{x}}t\right)\right]\right\} \exp\left(-\frac{x^2 + y^2}{2b_{\mathrm{g}}^2}\right)\,\mathrm{d}x\,\mathrm{d}y$$

$$= I_1 + I_2\;, \tag{H.11}$$

where

$$I_1 = \frac{I_0}{2} \int_{-\infty}^{\infty} \exp\left(-\frac{x^2}{2b_{\mathrm{g}}^2}\right)\,\mathrm{d}x \cdot \int_{-\infty}^{\infty} \exp\left(-\frac{y^2}{2b_{\mathrm{g}}^2}\right)\,\mathrm{d}y$$

$$= I_0\pi b_{\mathrm{g}}^2\;, \tag{H.12}$$

$$I_2 = \frac{I_0}{2} \int_{-\infty}^{\infty} \cos\left[\frac{2\pi}{p}\left(x - v_{\mathrm{x}}t\right)\right] \exp\left(-\frac{x^2}{2b_{\mathrm{g}}^2}\right)\,\mathrm{d}x \cdot \int_{-\infty}^{\infty} \exp\left(-\frac{y^2}{2b_{\mathrm{g}}^2}\right)\,\mathrm{d}y$$

$$= \frac{I_0 b_{\mathrm{g}} \sqrt{\pi}}{\sqrt{2}} \int_{-\infty}^{\infty} \cos\left[\frac{2\pi}{p}\left(x - v_{\mathrm{x}}t\right)\right] \exp\left(-\frac{x^2}{2b_{\mathrm{g}}^2}\right)\,\mathrm{d}x$$

$$= \frac{I_0 b_{\mathrm{g}} \sqrt{\pi}}{\sqrt{2}} \cdot \frac{1}{2} \exp\left(\mathrm{i}\frac{2\pi v_{\mathrm{x}}}{p}t\right) \int_{-\infty}^{\infty} \exp\left(-\frac{x^2}{2b_{\mathrm{g}}^2}\right) \exp\left(-\mathrm{i}\frac{2\pi}{p}x\right)\,\mathrm{d}x$$

$$+ \frac{I_0 b_{\mathrm{g}} \sqrt{\pi}}{\sqrt{2}} \cdot \frac{1}{2} \exp\left(-\mathrm{i}\frac{2\pi v_{\mathrm{x}}}{p}t\right) \int_{-\infty}^{\infty} \exp\left(-\frac{x^2}{2b_{\mathrm{g}}^2}\right) \exp\left(\mathrm{i}\frac{2\pi}{p}x\right)\,\mathrm{d}x$$

$$= \frac{I_0 b_{\mathrm{g}} \sqrt{\pi}}{\sqrt{2}} \left[\frac{1}{2} \exp\left(\mathrm{i}\frac{2\pi v_{\mathrm{x}}}{p}t\right) + \frac{1}{2} \exp\left(-\mathrm{i}\frac{2\pi v_{\mathrm{x}}}{p}t\right)\right]$$

$$\cdot \sqrt{2\pi}\, b_{\mathrm{g}} \exp\left(-\frac{2\pi^2 b_{\mathrm{g}}^2}{p^2}\right)$$

$$= I_0\pi b_{\mathrm{g}}^2 \cos\left(\frac{2\pi v_{\mathrm{x}}}{p}t\right) \cdot \exp\left(-\frac{2\pi^2 b_{\mathrm{g}}^2}{p^2}\right)\;. \tag{H.13}$$

Finally, the output signal is obtained as

$$g\left(t\right) = I_0\pi b_{\mathrm{g}}^2 \left[1 + \exp\left(-\frac{2\pi^2 b_{\mathrm{g}}^2}{p^2}\right) \cos\left(\frac{2\pi v_{\mathrm{x}}}{p}t\right)\right]\;. \tag{H.14}$$

By applying the above procedure to the rectangular-transmittace filter given by (H.9), (2.61) can be obtained.

References

1. F. Durst, A. Melling, J.H. Whitelaw: *Principles and Practice of Laser-Doppler Anemometry* (Academic Press, London 1976)
2. T.S. Durrani, C.A. Greated: *Laser Systems in Flow Measurement* (Plenum Press, New York 1977)
3. M. Raffel, C. Willert, J. Kompenhans: *Particle Image Velocimetry* (Springer, Berlin Heidelberg 1998)
4. B.M. Watrasiewicz, M.J. Rudd: *Laser Doppler Measurements* (Butterworths, London 1976)
5. L.E. Drain: *The Laser Doppler Technique* (John Wiley & Sons, Chichester 1980)
6. H.-E. Albrecht, M. Borys, N. Damaschke, C. Tropea: *Laser Doppler and Phase Doppler Measurement Techniques* (Springer, Berlin Heidelberg 2003)
7. T. Asakura, N. Takai: Appl. Phys. **25**, 179 (1981)
8. A.E. Ennos: 'Speckle Interferometry'. In: *Laser Speckle and Related Phenomena*. Ed. by J.C. Danity (Springer, Berlin Heidelberg 1984) pp. 203–253
9. H.E. Meinema: Electronics **26**, 135 (1953)
10. G.F. Aroyan: Proc. IRE **47**, 1561 (1959)
11. A.R. Gédance: J. Opt. Soc. Am. **51**, 1127 (1961)
12. J.T. Ator: J. Opt. Soc. Am. **53**, 1416 (1963)
13. J.T. Ator: Appl. Opt. **5**, 1325 (1966)
14. M. Gaster: J. Fluid Mech. **20**, 183 (1964)
15. M. Naito, Y. Ohkami, A. Kobayashi: J. Soc. Instrum. Control Eng. **7**, 761 (1968), (in Japanese)
16. S. Tsutsumi: Trans. Inst. Electron. Commun. Eng. **51-C**, 66 (1968), (in Japanese)
17. A. Kobayashi, M. Naito: Trans. Soc. Instrum. Control Eng. **5**, 142 (1969), (in Japanese)
18. M. Naito, M. Ishigami, A. Kobayashi: 'Spatial Filter and Its Application to Industrial Measurement'. In: *Proc. 5th IMEKO, 1970*, p. JA-129
19. A. Kobayashi, M. Nakayama, T. Yamaura, Y. Ohkami: Trans. Soc. Instrum. Control Eng. **15**, 89 (1979), (in Japanese)
20. S. Tsutsumi: OYO BUTURI – J. Jpn. Soc. Appl. Phys. **43**, 824 (1974), (in Japanese)
21. Y. Itakura, A. Sugimura, S. Tsutsumi: Appl. Opt. **20**, 2819 (1981)

22. M. Tsudagawa, S. Tsutsumi, H. Yamada: Trans. Inst. Electron. Commun. Eng. **64-C**, 843 (1981), (in Japanese)
23. T. Ushizaka, T. Asakura: Appl. Opt. **22**, 1870 (1983)
24. T. Koyama, J. Nitta, T. Asakura, T. Ushizaka, Y. Aizu, Y. Kikuchi: Biorheology, Suppl. I, 131 (1984)
25. Y. Aizu, T. Ushizaka, T. Asakura: Appl. Opt. **24**, 627 (1985)
26. Y. Aizu, T. Ushizaka, T. Asakura: Appl. Opt. **24**, 636 (1985)
27. Y. Aizu, T. Ushizaka, T. Asakura: Optik **71**, 55 (1985)
28. Y. Aizu, T. Ushizaka, T. Asakura, T. Koyama: Appl. Opt. **25**, 31 (1986)
29. J.L. Borders, H.J. Granger: Microvasc. Res. **27**, 117 (1984)
30. B. Reuter, M. Kratzer: 'Microanalysis of The Velocity Profile in Blood Vessel Models'. In: *Laser 79, Opto-Electronics Conference at Munich, Germany, July 2–6, 1979*, Ed. by W. Waidelich (IPC Science and Technology Press 1979) pp. 209–215
31. E. Delatour, M. Hanss: Rev. Sci. Instrum. **55**, 508 (1984)
32. A. Hayashi, Y. Kitagawa: Appl. Opt. **21**, 1394 (1982)
33. A. Hayashi, Y. Kitagawa: Trans. Inst. Electron. Commun. Eng. **66-C**, 717 (1983), (in Japanese)
34. A. Hayashi, J. Taguchi, Y. Kitagawa: Trans. Inst. Electron. Commun. Eng. **67-C**, 239 (1984), (in Japanese)
35. Y. Kitagawa, A. Hayashi : Trans. Inst. Electron. Commun. Eng. **68-C**, 505 (1985), (in Japanese)
36. W. Mitsuhashi, H. Mochizuki : Trans. Inst. Electron. Commun. Eng. **65-C**, 578 (1982), (in Japanese)
37. W. Röckemann, G.J. Plesse: Bibl. Anat. **11**, 50 (1973)
38. D.W. Slaaf, J.P.S.M. Rood, G.J. Tangelder, T. Arts: Microvasc. Res. **17**, S173 (1979)
39. D.W. Slaaf, J.P.S.M. Rood, G.J. Tangelder, T.J.M. Jeurens, R. Alewunse, R.S. Reneman, T. Arts: Microvasc. Res. **22**, 110 (1981)
40. B. Reuter, N. Talukder: 'New Differential Laser Microanemometer'. In: *1980 European Conference on Optical Systems and Applications at Utrecht, The Netherlands, September 23–25, 1980*, Proc. SPIE **236**, (SPIE 1981) pp. 226–230
41. H. Kiesewetter, H. Radtke, N. Körber, H. Schmid-Schönbein: Microvasc. Res. **23**, 56 (1982)
42. T. Ushizaka, Y. Aizu, T. Asakura: Appl. Phys. B**39**, 97 (1986)
43. E.A. Ballik, J.H.C. Chan: Appl. Opt. **12**, 2607 (1973)
44. J.H.C. Chan, E.A. Ballik : Appl. Opt. **14**, 1839 (1975)
45. E.A. Ballik, J.H.C. Chan: Appl. Opt. **16**, 674 (1977)
46. Y. Aizu, T. Ushizaka, T. Asakura: Appl. Opt. **24**, 641 (1985)
47. H. Ohno, T. Yamaura, A. Kobayashi: 'Spatial Filter Detector with Two-Dimensional M-Sequence Pattern and Its Application to Macro-Measurement of Random Movement'. In: *The 4th Sensor Symposium, Japan 1984*, (The Institute of Electrical Engineers of Japan, Tokyo 1984) pp. 61–66
48. F. Kobayashi, N. Chikahisa, A. Kobayashi: Trans. Soc. Instrum. Control Eng. **21**, 157 (1985), (in Japanese)
49. J.C.F. Wang, D.A. Tichenor: Appl. Opt. **20**, 1367 (1981)
50. K. Takemura, S. Tsutsumi: Trans. Inst. Electron. Commun. Eng. **64-C**, 582 (1981), (in Japanese)
51. S. Tsutsumi: Jpn. J. Soc. Mech. Eng. **82**, 1143 (1979), (in Japanese)

52. A. Kobayashi: J. Soc. Instrum. Control Eng. **19**, 409, 571 (1980), (in Japanese)
53. A. Papoulis: *Random Variables and Stochastic Process* (McGraw-Hill, New York 1965)
54. Y.W. Lee: *Statistical Theory of Communication* (John Wiley & Sons, New York 1960)
55. J.W. Goodman: *Introduction to Fourier Optics* (McGraw-Hill, San Francisco 1968)
56. Y. Aizu, T. Asakura: Appl. Phys. B **43**, 209 (1987)
57. H. Mishina, Y. Kawase, T. Asakura: Jpn. J. Appl. Phys. **15**, 633 (1976)
58. Y. Kawase, H. Mishina, T. Asakura: Jpn. J. Appl. Phys. **15**, 2173 (1976)
59. T. Sato, T. Kishimoto, K. Sasaki: Appl. Opt. **17**, 230 (1978)
60. O. Sasaki, T. Sato, T. Oda: Appl. Opt. **19**, 2565 (1980)
61. H.C. van de Hulst: *Light Scattering by Small Particles* (Dover, New York 1981)
62. K. Michel: *Ein Beitrag zur Signalverarbeitung von Ortsfiltersensoren* (VDI Verlag, Düsseldorf 2001)
63. G. Stavis: Instrum. Control Syst. **39**, 99 (1966)
64. T. Asakura: 'Surface Roughness Measurement'. In: *Speckle Metrology*. Ed. by R.K. Erf (Academic Press, New York 1978) pp. 11–49
65. L.M. Veselov, I.A. Popov: Opt. Spectrosc. **84**, 268 (1998)
66. M. Born, E. Wolf: *Principles of Optics*, 5th edn. (Pergamon, Oxford 1975)
67. R.J. Adrian: 'Limiting Resolution of Particle Image Velocimetry for turbulent flow'. In: *Advances in Turbulence Research – 1995, 2nd Turbulence Research Association Conference at Pohang Inst. Tech., 1995*
68. R.B. Johnson: 'Lenses'. In: *Handbook of Optics II*. 2nd edn., Ed. by M. Bass (McGraw-Hill, New York 1995) pp. 1.3–1.41
69. M. Gu: *Advanced Optical Imaging Theory* (Springer, Berlin 2000)
70. C. Berger: Opt. Eng. **41**, 2599 (2002)
71. W. Lauterborn, T. Kurz, M. Wiesenfeldt: *Coherent Optics* (Springer, Berlin 1995)
72. H. Mishina, S. Tokui, T. Ushizaka, T. Asakura, S. Nagai: Jpn. J. Appl. Phys. Suppl. **14**, 323 (1975)
73. H. Mishina, T. Asakura: Appl. Phys. **5**, 351 (1975)
74. E.O. Brigham: *The Fast Fourier Transform* (Prentice-Hall, New Jersey 1974)
75. R.N. Bracewell: *The Fourier Transform and Its Applications* International Student Edition (McGraw-Hill Kogakusha, Tokyo 1978)
76. J.P. Burg: 'Maximum Entropy Spectral Analysis'. In: *The 37th Annual International Meeting, Soc. Exploration Geophysicists, Oklahoma, October 31, 1967*
77. C.H. Chen: *Nonlinear Maximum Entropy Spectral Analysis Methods for Signal Recognition* (Research Studies Press, Chichester 1982)
78. H. Akaike: Annals of Inst. Statist. Math. **21**, 407 (1969)
79. T.J. Ulrych, T.N. Bishop: Rev. Geophys. and Space Phys. **13**, 183 (1975)
80. P.F. Fougere: J. Geophys. Res. **82**, 1051 (1977)
81. S. Haykin: *Nonlinear Methods of Spectral Analysis* (Springer, Berlin 1979)
82. T.S. Durrani, T.H. Wilmshurst, C. Greated: Opt-electronics **5**, 71 (1973)
83. T.H. Wilmshurst, J.E. Rizzo: J. Phys. E: Sci. Instrum. **7**, 924 (1974)
84. P. Venkatesh: J. Phys. E: Sci. Instrum. **14**, 587 (1981)
85. D.B. Brayton, H.T. Kalb, F.L. Crosswy: Appl. Opt. **12**, 1145 (1973)
86. H. Mishina, T. Asakura, S. Nagai: Opt. Commun. **11**, 99 (1974)

87. H. Mishina, T. Ushizaka, T. Asakura: Opt. Laser Technol. **8**, 121 (1976)
88. R.J. Adrian: J. Phys. E: Sci. Instrum. **5**, 91 (1972)
89. H. Mishina, T. Asakura: Appl. Phys. **8**, 179 (1975)
90. R.J. Adrian, J. Evensted, L.M. Fingerson, R. Menon: J. Phys. E: Sci. Instrum. **21**, 743 (1988)
91. T. Nakajima, Y. Ikeda: Meas. Sci. Technol. **1**, 776 (1990)
92. H.Z. Cummins, E.R. Pike: *Photon Correlation Spectrometry and Velocimetry* (Plenum Press, New York 1977)
93. R.G.W. Brown, K.D. Ridley, J.G. Rarity: Appl. Opt. **25**, 4122 (1986)
94. R.G.W. Brown, R. Jones, J.G. Rarity, K.D. Ridley: Appl. Opt. **26**, 2383 (1987)
95. R.G.W. Brown, M. Daniels: Appl. Opt. **28**, 4616 (1989)
96. R.J. McIntyre: Measurement **3**, 146 (1985)
97. H.Z. Cummins, E.R. Pike: *Photon Correlation and Light Beating Spectroscopy* (Plenum Press, New York 1974)
98. E.R. Pike, J.B. Abbiss: *Light Scattering and Photon Correlation Spectroscopy* (Kluwer Academic, Dordrecht 1997)
99. F. Francini, G. Longobanti: Rev. Sci. Instrum. **58**, 1106 (1987)
100. D. Malacara: 'Optical Testing'. In: *Handbook of Optics II*. 2nd edn., Ed. by M. Bass (McGraw-Hill, New York 1995) pp. 30.1–30.26
101. *Edmund Industrial Optics Catalog*. J034 (Edmund Optics Japan, Tokyo, 2003)
102. Y.C. Agrawal: J. Phys. E: Sci. Instrum. **17**, 458 (1984)
103. H. Inaba, Y. Itakura, M. Kasahara: Trans. Soc. Instrum. Control Eng. **29**, 1232 (1993), (in Japanese)
104. J. Ritonga, T. Ushizaka, T. Asakura: Appl. Phys. B **48**, 371 (1989)
105. J. Ritonga, T. Ushizaka, T. Asakura: J. Opt. (Paris) **20**, 145 (1989)
106. J. Ritonga, T. Ushizaka, T. Asakura: Opt. Quantum Electron. **22**, 143 (1990)
107. A. Kobayashi: OYO BUTURI – J. Jpn. Soc. Appl. Phys. **52**, 1007 (1983), (in Japanese)
108. S. Inagaki, T. Yamaura, A. Kobayashi: Trans. Soc. Instrum. Control Eng. **23**, 633 (1987), (in Japanese)
109. Y. Agrawal, J.B. Riley: Appl. Opt. **23**, 57 (1984)
110. Y. Amari, I. Masuda: IEEE Trans. Instrum. Meas. **39**, 649 (1990)
111. H. Ogiwara, H. Ukita: Jpn. J. Appl. Phys. Suppl. **14**, 307 (1975)
112. U. Schnell, J. Piot, R. Dändliker: J. Opt. Soc. Am. A **15**, 207 (1998)
113. K. Michel, O. Fiedler, A. Richter, K. Christofori, S. Bergeler: 'A Novel Spatial Filtering Velocimeter Based on a Photodetector Array'. In: *IMTC/97, 14th IEEE Instrum. Meas. Tech. Conf. at Ottawa, Canada, 1997* **2** (IEEE, New Jersey 1997) pp. 874–878
114. K. Michel, O. Fiedler, A. Richter, K. Christofori, S. Bergeler: IEEE Trans. Instrum. Meas. **47**, 299 (1998)
115. K. Christofori, K. Michel: Flow Meas. Instrum. **7**, 265 (1996)
116. H. Yamasaki, K. Oka, W. Mitsuhashi: 'An Adaptive Intelligent Velocity Sensing System'. In: *TRANSDUCERS' 87, 4th International Conference on Solid-State Sensors and Actuators at Tokyo, Japan, June, 1987* (The Institute of Electrical Engineers of Japan, Tokyo 1987) pp. 137–142
117. K. Oka, W. Mitsuhashi, H. Yamasaki: Trans. Soc. Instrum. Control Eng. **25**, 271 (1989), (in Japanese)
118. M. Terashima, M. Nomura, M. Hori, J. Shimomura: Trans. Inst. Elect. Eng. Jpn. **112-D**, 1056 (1992), (in Japanese)

119. K. Oka: 'Adaptive Sensor System'. In: *Intelligent Sensors*. Ed. by H. Yamasaki (Elsevier, Amsterdam 1996) pp. 99–108

120. R. Jain, S.L. Bartlett, N. O'brien: IEEE Trans. Pattern Analysis Machine Intelligence PAMI-**9-3**, 356 (1987)

121. W. Mitsuhashi, K. Oka, H. Yamasaki: Trans. Soc. Instrum. Control Eng. **24**, 1111 (1988), (in Japanese)

122. J. Czarske, F. Hock, H. Müller: 'Quadrature Demodulation – A New LDV-Burst Signal Frequency Estimator'. In:*5th International Conference on Laser Anemometry – Advances and Applications, 1993*, Proc. SPIE **2052**, (SPIE 1993) pp. 79–86

123. K. Michel, S. Bergeler, J. Kumpart, K. Christofori, H. Krambeer: 'High Resolution Velocity Measurements on Accelerating Surfaces by Means of a Spatial Filtering Sensor'. In: *SENSOR '99 at Nürnberg, Germany, 1999* **2** pp. 327–332

124. H. Miike, K. Koga, M. Momota, H. Hashimoto: Jpn. J. Appl. Phys. **26**, L1431 (1987)

125. H. Yamamoto, M. Momota, K. Koga, Miike: Trans. Inst. Electron. Information Commun. Eng. **75-D-II**, 1682 (1992), (in Japanese)

126. M.S. Uddin, H. Inaba, Y. Itakura, M. Kasahara: Appl. Opt. **37**, 6234 (1998)

127. M.S. Uddin, H. Inaba, Y. Itakura, Y. Yoshida, M. Kasahara: Appl. Opt. **38**, 6714 (1999)

128. Y. Itakura: 'Spatial-Filtering Velocimetry'. In: *Encyclopedia of Optical Engineering*. Ed. by R.G. Driggers (Marcel Dekker, New York 2003) pp. 2607-2617

129. S. Bergeler, H. Krambeer, K. Michel, D. Petrak: 'Velocity and Flow Rate Measurements in Capillaries'. In: *SENSOR 2001 at Nürnberg, Germany, 2001* **2** pp. 263–267

130. I. Menn, K. Michel, E. Dörp, H. Krambeer: Clin. Hemorheol. **24**, 122 (2001)

131. Y. Morikawa, Y. Tsuji, T. Tanaka: Bull. Jpn. Soc. Mech. Eng. **29**, 802 (1986)

132. D. Petrak, M. Köhler, K. Rosenfeld, E. Przybilla, S. Dietrich: Flow Meas. Instrum. **7**, 231 (1996)

133. O. Fiedler, N. Labahn, J. Kumpart, K. Christofori: 'Usage of The Spatial Filter Method for Measurements of Local Particle Velocities in Circulating Fluidized Beds'. In: *8th International Symp. on Application of Laser Techniques to Fluid Mech. at Lisbon, Portugal, 1996* (Radoan, Lisbon 1996) paper 12.1, pp. 1–8

134. Y. Sakai, T. Uno, J. Takagi, T. Yamashita: Opt. Rev. **2**, 65 (1995)

135. N. Shirai: 'Non-Contact Vehicle Speed Measurement Using Spatial Filter'. In: *Journal of Mechanical Engineering Laboratory.* **32** (1978) pp. 292–299, (in Japanese)

136. OMRON Product Catalog *Ground View Sensor*. Catalog ①ZN①018 (Omron Corp., Tokyo, 2001)

137. Y. Murata, K. Egawa, Y. Shinmoto: 'Road Surface Recognition Sensor Using an Optical Spatial Filter'. In: *Omron Technics.* **37** (Omron Corp., Kyoto 1997) pp. 134–139, (in Japanese)

138. Y. Shinmoto, J. Takagi, K. Egawa, Y. Murata, M. Takeuchi: 'Road Surface Recognition Sensor Using an Optical Spatial Filter'. In: *IEEE International Conference on Intelligent Transportation System,* (1994) pp. 1000–1004

139. ONO SOKKI Product Catalog *Speedmeter LC Series*. Catalog 414-11 (Ono Sokki, Kanagawa, Japan 2000)

140. S.W. Li, T. Aruga: Appl. Opt. **31**, 560 (1992)

141. S. Komatsu, I. Yamaguchi, H. Saito: Opt. Commun. **18**, 314 (1976)

142. A. Hayashi, Y. Kitagawa: Opt. Commun. **43**, 161 (1982)
143. A. Hayashi, Y. Kitagawa: Opt. Commun. **49**, 91 (1984)
144. A. Hayashi, Y. Kitagawa: Trans. Inst. Electron. Commun. Eng. **67-C**, 33 (1984), (in Japanese)
145. Y. Kitagawa, A. Hayashi : Appl. Opt. **24**, 955 (1985)
146. M.L. Jakobsen, S.G. Hanson : Appl. Opt. **43**, 4643 (2004)
147. M.L. Jakobsen, S.G. Hanson : Meas. Sci. Technol. **15**, 1949 (2004)
148. J. Ritonga, T. Ushizaka, T. Asakura: J. Opt. (Paris) **21**, 9 (1990)
149. W.M. Farmer: Appl. Opt. **11**, 2603 (1972)
150. R.J. Adrian, K.L. Orloff: Appl. Opt. **16**, 677 (1977)
151. D.V. Semenov, E. Nippolainen, A.A. Kamshilin : Opt. Lett. **30**, 248 (2005)
152. I. Yamaguchi, T. Furukawa, T. Ueda, E. Ogita: Opt. Eng. **25**, 671 (1986)
153. I. Yamaguchi, T. Okamoto, H. Nagayama: Meas. Sci. Technol. **1**, 406 (1990)
154. T. Murakami, I. Yamaguchi: Kogaku – Jpn. J. Opt. **17**, 239 (1988), (in Japanese)
155. M.L. Jakobsen, H.E. Larsen, S.G. Hanson : J. Opt. A: Pure Appl. Opt. **7**, S303 (2005)
156. R. Semiat, A.E. Dukler: AIChE J. **27**, 148 (1981)
157. A. Cartellier: Appl. Opt. **25**, 2815 (1986)
158. A. Cartellier: Appl. Opt. **31**, 3493 (1992)
159. J.H.C. Chan, E.A. Ballik : Appl. Opt. **13**, 234 (1974)
160. S.W. Li, T. Aruga: Appl. Opt. **32**, 2320 (1993)
161. H. Kadono, R. Uma Maheswari, N. Takai, T. Asakura: Opt. Commun. **69**, 199 (1989)
162. H. Kadono, S. Toyooka: Opt. Commun. **74**, 159 (1989)
163. Y.Y. Hung: 'Displacement and Strain Measurement'. In: *Speckle Metrology*. Ed. by R.K. Erf (Academic Press, New York 1978) pp. 51–71
164. M. Azzazy, R.L. Potts, L. Zhou, B. Rosow: Appl. Opt. **36**, 2721 (1997)
165. Y. Aizu, K. Ogino, T. Asakura: Opt. Commun. **64**, 205 (1987)
166. L.J. Weng: IEEE Trans. Information Theory **17**, 457 (1971)
167. F.J. McWilliams, N.J.A. Sloane: Proc. IEEE **64**, 1715 (1976)
168. E. Hecht: *Optics*, 4th edn. (Addison-Wesley, Massachusetts 2002)

Index

Springer Series in
OPTICAL SCIENCES

Springer Series in
OPTICAL SCIENCES